Current Topics in Microbiology 146 and Immunology

New Strategies for Oral Immunization

International Symposium
at the University of Alabama at Birmingham
and Molecular Engineering Associates, Inc.
Birmingham, AL, USA, March 21-22, 1988

Organized and Edited by
J. Mestecky, and J. R. McGhee

With 22 Figures

Springer-Verlag
Berlin Heidelberg NewYork
London Paris Tokyo

JIRI MESTECKY, M.D.

Professor of Microbiology
Medicine and Oral Biology

JERRY R. MCGHEE, PH. D.

Professor of Microbiology

Department of Microbiology,
University of Alabama at Birmingham
UAB Station, Birmingham, Alabama 35 294, USA

ISBN-13: 978-3-642-74531-7 e-ISBN-13: 978-3-642-74529-4
DOI: 10.1007/978-3-642-74529-4

Library of Congress Catalog Card Number 15-12910

Softcover reprint of the hardcover 1st edition 1989

Offsetprinting: Color-Druck Dorfi GmbH, Berlin

2123/3020-543210 – Printed on acid-free paper

Preface

Environmental antigens, particularly those from ingested food and endogenous gastrointestinal microorganisms, are important stimulants of the hosts immune system. Thus, both humoral and cell-mediated arms of the systemic and mucosal immune compartments are profoundly influenced by the constant exposure to such environmental stimuli. It might be surprising to some to realize that the idea of stimulating protective immunity by oral immunization was actually discovered by a pioneering scientist and forebearer of modern immunology. Paul Ehrlich showed in two largely neglected papers published in 1891, that the oral route of immunization has distinct advantages over systemic immunization for protective immunity to the plant poisons abrin and ricin. Even more remarkable was his demonstration that such immunization protected guinea pigs to not only a systemic but also to a mucosal challenge. He found that **oral immunization** with increasing doses of toxin protected animals from **conjunctival challenge**, while control animals developed massive necrotic lesions in the challenged eye. It should be stressed that these results were reported only a few years after the discovery of antibodies and decades before it was realized that different immunoglobulin isotypes occur and are found in characteristic distribution in various external secretions. Since Ehrlich induced protective immune responses at distant mucosal sites following oral immunization, he should be credited as the first to discover the common mucosal immune system.

Oral immunization has continued to have a fascinating history. It is difficult to select an antigen which has not been used by the oral route for immunization. In spite of the frequently successful use of this route to induce immunity, it is most often overshadowed by reports stressing other, usually systemic, routes for immunization. Thus, only a few of the numerous vaccines in current use are given by the oral route. We, therefore, felt that this neglect is not deserved and that oral immunization has much to contribute, particularly, in light of recent methodological and technical advances in antigen delivery systems, to modern human and veterinary preventive medicine.

Realization of this meeting was possible due to the support provided by numerous contributors: Molecular Engineering Associates, Birmingham, Alabama; the Procter and Gamble Company, Cincinnati, Ohio; and numerous colleagues at The University of Alabama at Birmingham, namely, Drs. J. Claude Bennett, C.O. Elson, G.H. Cassell, and Deans J. Pittman and R.R. Ranney from the Schools of Medicine and Dentistry.

It gives us a great deal of pleasure to acknowledge the help of Ms. Maria Bethune for the organization of this meeting as well as the typing and assembly of all manuscripts.

Jiri Mestecky, Jerry R. McGhee

Ehrlich P (1891) VI. Experimentelle Untersuchungen über Immunität. I. Ueber Ricin. Deutsche Medicinische Wochenschrift 17:976-979.
Ehrlich P (1891) Experimentelle Untersuchungen über Immunität. II. Ueber Abrin. Deutsche Medicinische Wochenschrift 17:1218-1219.

Contents

Oral Immunization and Mucosal Immune System

Antigen Delivery Systems

Viral Vaccines

Bacterial Vaccines

List of Contributors

You will find the adresses at the beginning of the respective contribution

Oral Immunization and Mucosal Immune System

Oral Immunization: Past and Present

J. Mestecky, and J.R. McGhee

Departments of Microbiology and Medicine, University of Alabama at Birmingham, Birmingham, AL 35294, USA

INTRODUCTION

Diverse environmental antigens of microbial and food origin constantly stimulate the entire immune system. Mucosal membranes which represent a vast area of contact with the environment, are exposed daily to antigenic substances that induce specific humoral as well as cell-mediated immune responses not only at the site of the stimulation - mucosa-associated lymphoid tissues - but also in the draining lymph nodes, spleen, and the bone marrow (Fig. 1). Therefore, it should not be surprising that mucosa-associated organs, especially the intestines, contain the largest accumulation of lymphoid cells, including B and T lymphocytes and plasma cells, as well as accessory and antigen- processing and-presenting cells (Brandtzaeg this volume). Quantitatively, the intestinal tract is the most active organ engaged in immunoglobulin production and antibodies, particularly those of the IgA isotype, are selectively transported into the gut lumen (Conley and Delacroix 1987; Mestecky and McGhee 1987).

Mucosal surfaces are the most frequent portals of entry of common viral, bacterial, and parasitic infectious diseases and of potentially harmful antigenic substances from the environment. It has been convincingly demonstrated that secretory antibodies are able to limit the absorption of protein antigens through the mucosal membranes, inhibit the attachment of bacteria to the epithelial cells, and neutralize a broad spectrum of viruses that infect epithelial cells, or cause more generalized diseases (Kilian et al. 1988).

Because the currently used systemic immunization may not be effective in the induction of antibodies in external secretions, especially in individuals who had not been previously exposed to a given microorganism or environmental antigen through mucosal membranes, numerous attempts have been made to induce an immune response in external secretions (for review see Mestecky 1987). Thus, mucosal surfaces of the upper respiratory tract, conjunctiva, vagina, or intestinal tract have been exposed to killed or attenuated viruses or bacteria in many attempts to induce, frequently with success, a *local* immune response and protection of mucosal areas restricted to the site of antigen application. Subsequent extensive studies that addressed the questions concerning the origin of IgA-producing plasma cells and effective routes of immunization revealed that gut- and bronchus-associated-lymphoid tissues function as sources of antigen-sensitized and IgA-committed plasma cell precursors that populate remote mucosal tissues and glands and led to the formulation of the concept of a common mucosal immune system (for reviews see Lamm 1976; Mestecky and McGhee 1987).

In addition to humoral responses manifested by secretory antibodies, antigen ingestion profoundly influences systemic immunity (Fig. 1). Of particular interest for the prevention of sensitization by environmental allergens is the state of oral tolerance (Tomasi 1980) that appears to be predominately mediated by a subset of T cells and their

Current Topics in Microbiology and Immunology, Vol. 146

products. Succinctly summarized, ingestion of conjugated haptens or antigens of food and microbial origin suppresses, in certain experimental models, an immune response to a subsequent systemic challenge with the same antigen. This originally empirical observation prompted extensive studies carried out mainly in the first half of this century with allergens of plant origin.

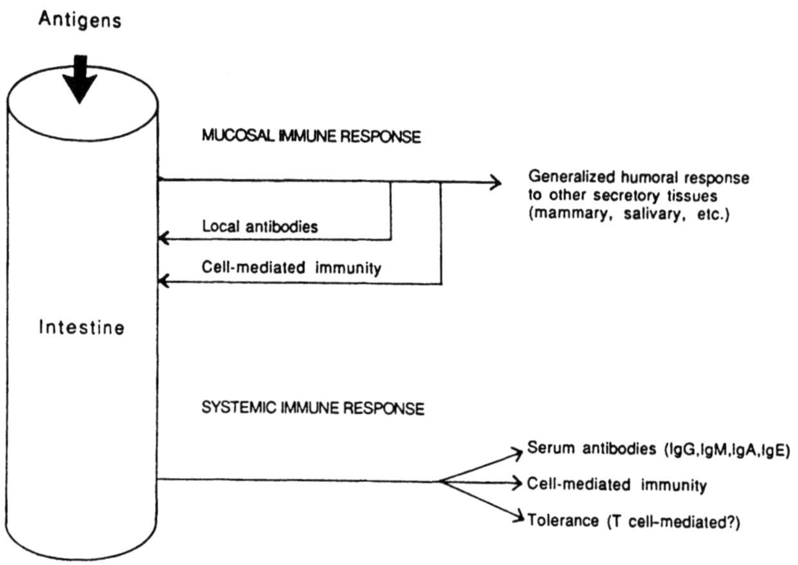

Fig. 1. Types of mucosal and systemic immune responses induced by intestinal exposure to antigens.

HISTORICAL ASPECTS

Close examination of the history of oral immunization and of the protective effects of induced secretory and serum antibodies or of systemic tolerance to potential allergens has revealed that a voluminous literature, currently neglected or misquoted, is available on this subject. It is disheartening to note that thorough studies performed at the end of the last and the beginning of this century have not been re-interpreted in the light of modern immunology. Time and space limitations will not permit us to provide a comprehensive review which this subject would deserve. Nevertheless, highlights of selected and frequently entertaining studies will clearly illustrate the chronicle of past events in this field.

Ingestion of foreign substances represents the first well documented attempts in the enhancement of resistance. In fear for his life, the king of Pontus (territory on the Northern shores of todays Turkey), Mithridates VI-Eupator (~ 132 - 63 B.C.), invented a universal antidote known by his name (Reinach 1890). The formula in his own handwriting was found in his archieves: "Take two dry nuts, two figs, twenty leaves of rue (an aromatic Eurasian plant of the genus *Ruta*; leaves contain volatile oil formerly used in medicine as "herb-of-grace"), crush artfully and add a pinch of salt. This potion taken on an empty stomach protects all day against the effects of poison". According to other sources, he mixed it with the blood of pontine ducks habitually fed poisonous weeds. This mixture was taken every morning. The king always carried a lethal dose of poison extracted from plants on his person, enclosed in the hilt of his scimitar. After the last battle with the Roman army under general Pompey and revolt of his own soldiers led by his son Farnaces II, Mithridates tried in vain to poison himself, although the poison from the same vial proved lethal for two of his daughters, Mithridatis and Nysa. His devoted Gallic mercenary was ordered to stab the desperate king. In modern interpretation, Mithridates VI Eupator is given credit by some (Witebsky 1967) for the first attempt to induce immunity. It is pleasing to note that this goal was achieved by the oral route.

Fig. 2. Mithridates VI, Eupator (~ 132-63 B.C.), King of Pontus.

It was Louis Pasteur (1822-1895) and his colleagues, Roux and Chamberland, who in 1880 reported in their studies of chicken cholera the first successful protection of the fowl, that was induced by the addition of *B. avisepticum* to chicken feed (for review see Gay 1924). The interest in oral immunization was retained in the Pasteur Institute by the pupils and followers of the great teacher. Specifically, Klemperer in 1892 and Metchnikoff in 1903 induced agglutinins in human and animal sera by oral immunization with killed or live *V. cholerae*. Calmette and Guerin published several papers from 1906 to 1923 in which serum antibodies were induced by oral immunization with *M. tuberculosis*, *Y. pestis*, or *C. diphtheriae* (summarized in Gay 1924). Parallel studies with *S. dysenteriae*, performed by Besredka at the Pasteur Institute, indicated that the ingestion of killed bacteria induced immunity in the absence of serum antibodies. His extensive studies were published in a classical monograph on local immunity (Besredka 1927). These observations are reminiscent of experiments performed by Bull and McKee (1929) who intranasally immunized rabbits with a killed pneumococcal vaccine before a challenge with a virulent pneumococcus. In comparison with control animals or those immunized with a single dose, two to four immunizations were sufficient to prevent fatal pneumonia and septicemia (Table

1). Interestingly, no serum antibodies were detectable by the complement-fixation test, and the authors concluded that the resistance to intranasal infection can be induced in the absence of demonstrable serum antibodies. However, these studies were performed some 30 years before the identification of IgA as another immunoglobulin class (Heremans et al. 1959) and thus it is very likely that serum and secretory IgA anti-pneumococcal antibodies were indeed induced but escaped detection because of their apparent inability to bind complement in a classical complement-fixation test.

Table 1. Protection achieved by repeated intranasal immunization of rabbits with pneumococcus (from Bull and McKee 1929; Table XIII).

No. of treatments.	Died.	Infected but recovered.	Escaped infection.
4	7.7 per cent	7.7 per cent	84.6 per cent
2	14.3 per cent	14.3 per cent	71.4 per cent
1 (11 days before infection)	0	83.4 per cent	16.6 per cent
1 (8 days before infection)	83.4 per cent	0	16.6 per cent
Normal rabbits (10 series of 6)	57 per cent	27.0 per cent	16.0 per cent

The above mentioned observations were first exploited in 1922 by Vaillant who compared the efficacy of the oral vs subcutaneous vaccine containing *S. typhi* (see in Gay 1924). In 1236 subjects vaccinated orally there were only two cases of typhoid (0.16%), in 173 subjects who were vaccinated subcutaneously there were four cases (2.3%), whereas among over six hundred unvaccinated individuals there were fifty cases (7.7%) of typhoid. Thus, protection superior to a subcutaneous vaccine was achieved by oral immunization. An admittedly incomplete list of other microorganisms that have been used in oral vaccines in veterinary and human medicine provides convincing evidence that confirms the feasibility of this approach (Mestecky 1987) (Table 2). Of greater interest are recent studies summarized elsewhere (Mestecky 1987) and reported in this symposium which indicate that secretory antibodies and, in some cases, protection can be induced by oral immunization of humans with *E. coli, V. cholerae, H. influenzae, S. mutans, S. typhi*, polio, influenza, adeno, rota, and rabies viruses.

ORAL IMMUNIZATION AND THE INHIBITION OF HYPERSENSITIVITY REACTIONS

Use of oral immunization for desensitization of allergic patients also has a surprisingly long history. The first well documented but unsuccessful attempt of desensitization by ingestion of poison oak leaves dates from 1829 (Dakin).

> ..."*The shrubs most virulent are the poison oak, (Rhus toxicodendron,) and the poison ivy. Some good meaning, mystical, marvellous physicians, or favoured ladies with knowledge inherent, say the bane will prove the best antidote, and hence advise the forbidden leaves to be eaten, both as a preventive and cure to the external disease. I have known the experiment tried, which resulted in an eruption, swelling, redness, and intolerable itching, around the verge of the anus.*" ...

Table 2. Selected list of microbial, food, and environmental antigens used in various oral immunization studies in humans and in animals.*

Antigen	Comment
Chicken cholera	Protection induced (Pasteur 1880)
V. cholerae	Moderate protection (Klemperer 1892; Metchnikoff 1903). Oral vaccine for human use now available (Holmgren this volume)
M. tuberculosis *Y. pestis* *C. diphtheriae*	Serum antibodies induced (Calmette and Guerin 1906-1923)
S. dysenteriae	Secretory and serum antibodies, some protection (Besredka 1919-1927)
S. typhi	Oral immunization preferable to systemic (Vaillant 1922) Oral vaccine now available (Levine 1987)
S. pneumoniae	Protection achieved by intranasal immunization (Bull and McKee 1929)
E. coli	Antibodies in milk (Goldblum et al. 1975)
S. mutans	S-IgA antibodies in saliva and tears (Mestecky et al. 1987)
H. influenzae	IgA antibodies in saliva (Clancy this volume)
Other bacteria: *S. aureus, S. typhimurium, P. multocida, B. pertussis, B. suis, C. jejuni*	
Viruses: Polio, Adeno, Influenza, Rota, Rabies, Newcastle disease, Fowl pox, Infectious laryngotracheitis, Avian encephalitis, Sendai, Marek's disease	Human and animal studies. Some oral vaccines now available
Allergens Type I (Anaphylaxis) and Type IV (Delayed-type) Hypersensitivity:	

Abrin, ricin	Decreased skin and mucosal sensitivity (Ehrlich 1891)
Proteins from rice, maize, oats, flour	Serum antibodies (Magnus 1908)
Cow's milk, ox blood, egg white, zein, oats	Decrease in systemic reactivity (Wells and Osborne 1911)
Dinitrochlorobenzene	Inhibition of skin reactivity (Chase 1946)
Birch pollen, ragweed, grass pollen	Serum and secretory antibodies, improved clinical symptoms (Bjorksten et al. 1986; Taudorf et al. 1987; Wheeler et al. 1987; Waldman and Bergmann 1987)
Poison ivy	No effect; others observed partial or complete protection (Dakin 1822; Schamberg 1925; Shelmire 1941; Zisserman 1941; Gold and Masucci 1942)

*Assembled from Bjorksten and Dewdney 1987; Bull and McKee 1929; Chase 1946; Dakin 1829; Gay 1924; Holmgren this volume, International symposium on vaccination of man and animals by the non-parenteral route 1976; Levine et al. 1987; Mestecky 1987; Platts-Mills 1987; Stevens 1945; Tomasi 1980; and Wells and Osborne 1911.

The interest in oral desensitization by plant extract was revived in this country at the end of the last century and lasted until the end of the Second World War (for review see Stevens 1945). In many studies, dermatologist treated their patients with alcohol, ether, or acetone extracts of poison ivy leaves diluted in corn oil and given as a solution or in the form of enterically coated capsules. Decreased susceptibility or even complete protection reported in some, but not all of the studies, was found to parallel diminished sensitivity induced by oral hyposensitization. Although the mucosal membranes, according to most observers, were surprisingly unaffected by the ingestion of poison ivy extracts, perianal rashes of various intensities were frequently observed. With the development of modern anti-inflammatory drugs, oral immunization is not considered as a means for the prevention of cell-mediated hypersensitivity reactions in humans.

Ehrlich (1891) was the first to report that feeding of animals with abrin or ricin reduces inflammatory reactions upon subcutaneous inoculation with these plant poisons. Furthermore, oral immunization with ricin protected animals from a necrotic dose of ricin introduced into the *conjunctival sac*. Thus, *in 1891 Ehrlich for the first time clearly demonstrated that oral immunizatin results in the protection of an anatomically remote mucosal site - conjunctiva.* Reassessment of this finding in light of modern immunology leads to an inescapable conclusion that it was Ehrlich who foretold the existence of the common mucosal immune system and demonstrated the protective effect of oral immunization. Analogous observations were reported by Wells and Osborne (1911) in animals that had been fed, for an extended time, cow's milk, ox blood, egg white, zein from corn, or oat proteins and subsequently challenged for the development of anaphylaxis. In 1946 Chase reported in his now classical study that the experimental drug (dinitrochlorobenzene) allergy can be effectively inhibited by prior feeding of the sensitizing agent. Currently, the oral immunization route is gaining a momentum for its potential to effectively inhibit antibody (IgE)-mediated hypersensitivity (Type I) reactions on mucosal membranes of the respiratory tract through the induction of competing secretory antibodies or possibly suppressor T cells involved in the regulation of IgE synthesis. In this context, promising initial results have been obtained in patients orally immunized with birch pollen, ragweed, or other allergens (Bjorksten and Dewdney 1987; Platts-Mills 1987). There is little doubt that many studies in this clinically important area will ensue.

ANTIGEN DELIVERY SYSTEMS

Although ingestion of particulate as well as soluble antigens containing proteins, glycoproteins, polysaccharides, or haptens with carriers results, in several experimental models, in the induction of humoral and cellular immune responses (and/or oral tolerance), the required doses of antigens are much higher when compared to doses used for systemic immunizations. This fact has been a major reason as to why the oral immunization route has not been widely used. Denaturation of highly purified antigens such as isolated microbial products (viral glycoproteins, bacterial enzymes) by gastric hydrochloric acid, digestion with proteolytic enzymes, and their limited absorption (Hemmings 1978; Walker 1986), which essentially represents only minute amounts of ingested antigens, as well as their interactions with nonspecific factors or preexisting specific antibodies on mucosal surfaces negatively influence the outcome of the desired immune response.

To avoid degradation, antigens have been administered with a solution of sodium bicarbonate or packaged in gelatin capsules coated with substances that are insoluble under acidic conditions. There is little doubt that recent technological advances in antigen packaging, boosting of an immune response with oral adjuvants, and genetic engineering of microorganisms will maximize the effectiveness of orally administered vaccines (for review see Mestecky 1987). Although the advantages and disadvantages of individual approaches will be critically discussed in detail in the last contribution, a brief evaluation of such methods foretells of promising expectations. The immunogenicity of soluble protein or glycoprotein antigens which are not highly effective by themselves may be significantly potentiated by chemical coupling to cholera toxin B subunit which binds with an exquisite specificity to ganglioside receptors on epithelial cells which, in turn, are likely to function as efficient antigen-presenting cells (Elson this volume; Holmgren this volume). Together with appropriate coating which will permit solubilization and release of such conjugated vaccines at the desired site of stimulation, this principle may be exploited in the design of several potential vaccines.

Because of their multiplication in the gastrointestinal tract and potential to selectively colonize gut-associated lymphoid tissues, live microorganisms are more potent inducers of an immune response than their killed counterparts. Thus attenuated, genetically modified *S. typhi* has been successfully used in an extensive field trial of an orally administered vaccine (Levine et al. 1987). Furthermore, the potential of avirulent strains of *S. typhimurium* to selectively colonize Peyer's patches has been exploited in novel vaccines in which genetically manipulated "carrier" bacteria express gene products of other microbial species. This intriguing approach (Clements 1987) has been used experimentally in recent studies (Curtiss 1988) and vaccines based on these principles may be applied particularly to short-term immunizations in animals in which a simultaneous protection would be provided against *Salmonellae* and desired bacterial or viral antigens expressed in this species. Alternative approaches that have been used to facilitate the uptake of soluble antigens and to enhance the immune response in the gut include the incorporation of such antigens into liposomes (Michalek this volume) or microcapsules (Eldridge this volume). Initial studies with orally administered liposomes containing antigens derived from *S. mutans* have been promising (Bruyere et al. 1987; Michalek this volume). Particularly exciting are recent experiments with polysaccharide, protein or viral antigens incorporated into microcapsules that are taken up by Peyer's patches. Depending on their size and the rate of solubilization, predominately mucosal or systemic, long lasting immune responses can be induced (Eldridge this volume). Although the revived interest in alternative immunization routes have prompted many recent investigations, we believe that the orally administered vaccines have not been exploited to their fullest potential. Yet, it will become obvious from this symposium that this strategy offers attractive possibilities with respect to the simplicity of administration of vaccines and stimulation of a specific immune response at the site of entry of infectious agents and allergens. Ensuing technological improvements including the potential use of synthetic peptides (Steward and Howard 1987) incorporated into suitable carriers will undoubtedly foster a greater acceptance of this immunization route.

REFERENCES

Besredka A (1927) Local immunization. Williams and Wilkins, Baltimore.

Bjorksten B, Dewdney JM (1987) Oral immunotherapy in allergy - is it effective? Clin Allergy 17:91-94

Bruyere T, Wachsmann D, Klein J-P, Scholler M, Frank RM (1987) Local response in rat to liposome-associated *Streptococcus mutans* polysaccharide-protein conjugate. Vaccine 5:39-42

Bull CG, McKee CM (1929) Respiratory immunity in rabbits. VII. Resistance to intranasal infection in the absence of demonstrable antibodies. Am J Hyg 9:490-499

Chase MW (1946) Inhibition of experimental drug allergy by prior feeding of the sensitizing agent. Proc Soc Exp Biol Med 61:257-259

Clements JD (1987) Use of attenuated mutants of *Salmonella* as carriers for delivery of heterologous antigens to the secretory immune system. Pathol Immunopathol Res 6:137-146

Conley ME, Delacroix DL (1987) Intravascular and mucosal immunoglobulin A: Two separate but related systems of immune defense? Ann Intern Med 106:892-899

Dakin R (1829) Remarks on a cutaneous affliction, produced by certain poisonous vegetables. Am J Med Sci 4:98-100

Ehrlich P (1891) Experimentelle Untersuchungen über Immunität. I. Ueber Ricin. Deutsch Med Wochenschr 17:976-979

Ehrlich P (1891) Experimentelle Untersuchungen über Immunitatät. II. Ueber Abrin. Deutsch Med Wochenschr 17:1218-1219

Gay FP (1924) Local resistance and local immunity to bacteria. Physiol Rev 4:191-214

Goldblum RM, Ahlstedt S, Carlsson B, Hanson LA, Jodal U, Lidin-Janson G, Sohl-Ackerlund A (1975) Antibody-forming cells in human colostrum after oral immunization. Nature 257:797-799

Hemmings WA (ed) (1978) Antigen absorption by the gut. University Park Press, Baltimore

Heremans JF, Heremans MT, Schultze HE (1959) Isolation and description of a few properties of the β_2A globulin of human serum. Clin Chim Acta 4:96-102

Internationl symposium on vaccination of man and animals by the non-parenteral route (1976). Develop Biol Standard 33, S. Karger, Basel

Kilian M, Mestecky J, Russell MW (1988) Defense mechanisms involving Fc-dependent functions of immunoglobulin A, and their subversion by bacterial IgA proteases. Microbiol Rev 52:296-303

Levine M, Black RE, Ferreccio C, Germanier R (1987) Large-scale field trial of Ty21A live oral typhoid vaccine in enteric-coated capsule formulation. Lancet i: 1049-1052

Mestecky J (1987) The common mucosal immune sytem and current strategies for induciton of immune response in external secretions. J Clin Immunol 7:265-276

Mestecky J, McGhee JR (1987) Immunoglobulin A (IgA): Molecular and cellular interactions involved in IgA biosynthesis and immune response. Adv Immunol 40: 153-245

Platts-Mills TAE (1987) Oral immunotherapy: A way forward? Allergy Clin Immunol 80:129-132

Reinach T (1890) Mithridate Eupator. Roi de Pont. Firmin-Didot, Paris, pp 283-285, 409-410

Stevens FA (1945) Council on pharmacy and chemistry. Report of the council: Status of poison ivy extracts. J Am Med Assoc 127:912-921

Steward MW, Howard CR (1987) Synthetic peptides: A next generation of vaccines? Immunol Today 8:51-58

Tomasi Jr TB (1980) Oral tolerance. Transplantation 29:353-356

Walker WA (1986) Allergen absorption in the intestine: Implications for food allergy in infants. J Allergy Clin Immunol 78:1003-1009

Wells HG, Osborne TB (1911) The biological reactions of the vegetable proteins. I. Anaphylaxis. J Infect Dis 8:66-124

Witebsky E (1967) Immunology: Views of the past and visions for the future. Canadian Med Assoc J 97:1371-1377

Overview of the Mucosal Immune System

P. Brandtzaeg

Laboratory for Immunohistochemistry and Immunopathology (LIIPAT), Institute of Pathology, University of Oslo, The National Hospital, Rikshopitalet, N-0027 Oslo 1, Norway

INTRODUCTION

The existence of a protective local immune system that seemed to function independently of systemic immunity, was initially proposed by Besredka (1919); he observed that rabbits, after oral immunization with killed Shiga bacillus, were protected against fatal dysentery irrespective of the serum antibody titer. Davies (1922) obtained evidence in soldiers stationed around Jerusalem, to suggest for the first time the presence of local immunity in humans; his observations indicated that bacterial agglutinins could appear in dysentery stools several days earlier than in the blood. The molecular basis for local immunity was established when Tomasi et al. (1965) confirmed that external secretions contained a unique immunoglobulin (Ig) subsequently called secretory IgA (SIgA).

NATURE OF THE MUCOSAL IMMUNE SYSTEM

The exocrine secretory tissues constitute the most important mediator organ of specific humoral immunity - with the gut as the major contributor. There are about 10^{10} Ig-producing cells per meter of human small bowel (Brandtzaeg et al. 1987a). This is an impressive figure compared with that determined for bone marrow, spleen, and lymph nodes collectively (2.5×10^{10}). A rough estimate would thus indicate that 70-80% of all Ig-producing cells in human are located in the intestinal mucosa. A comparable estimate (>80%) was recently reported for mice (van der Heijden et al. 1987).

Most of the immunocytes adjacent to exocrine glands (70-90%) form J chain-containing dimers or larger polymers of IgA (Brandtzaeg 1985). Such polymeric IgA (pIgA) can be transported through the glandular epithelium along with J-chain positive pentameric IgM (pIgM) via a pIg receptor called the secretory component or SC (Brandtzaeg 1985; Mestecky and McGhee 1987). More IgA is translocated in this way to the gut lumen every day than the total daily production of IgG (which is approximately 30 mg/kg), whereas human bile contributes only about 1 mg/kg/day to the luminal IgA pool (Delacroix 1985).

It is still obscure how the particular immunoregulatory requirements of local immunity are fulfilled. Although the concept of a "common mucosal immune system" has gained wide acceptance, there are several indications that regulatory mechanisms operating in gut-associated lymphoid tissue (GALT) and in organized lympho-epithelial structures of the upper respiratory tract, such as tonsils and bronchus-associated lymphoid tissue (BALT), differ in several ways (reviewed in Brandtzaeg et al. 1987a; Sminia et al. 1989).

Current Topics in Microbiology and Immunology, Vol. 146

DISTRIBUTION AND CHARACTERISTICS OF MUCOSAL LYMPHOCYTES

Migration of Lymphocytes

An important basis for local immunity is the migration of specific B and T cells from GALT and BALT to various secretory tissues such as the gut mucosa (Fig. 1). Local extravasation of lymphoid cells belonging to the mucosal immune system seems to depend primarily on selective endothelial recognition mechanisms (Streeter et al. 1988). However, experimental evidence indicates that the dissemination of precursor cells to distant sites is more limited from BALT than from GALT (Sminia et al. 1989).

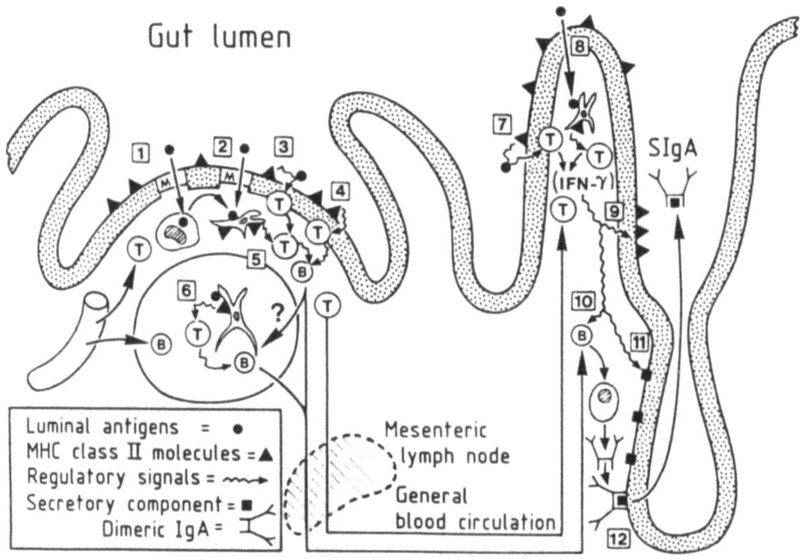

Fig. 1. Schematic illustration of possible sites of interaction between lymphoid cells (*T* and *B*), gut epithelium (*dotted*), and various accessory histiocytic cells. [1] Luminal antigen is transported into Peyer's patch through the M cell and is presented to T cells by subepithelial MHC class II-positive dendritic cells after being processed in the macrophage. [2] Antigen is transferred directly to the dendritic cell. [3] Antigen that has been processed in the gut lumen is presented to the T cell by class II-positive epithelial cell. [4] Autoreactive T cells are stimulated by a class II-positive epithelium to induce B-cell differentiation (e.g. switching to IgA expression). [5] Activated T cells provide various regulatory signals to the B cell. [6] Antigen presentation performed by follicular dendritic cell. T-cell activation and B-cell differentiation takes place in follicular germinal centers, but it is unknown (*?*) whether the intra- or extra-follicular compartment is more important as a source of stimulated B cells migrating to secretory tissues. [7] Antigen presentation by class II-positive villous epithelium affords further T-cell activation at a secretory site. [8] Such activation can likewise lead to absorption or penetration of antigen and presentation by the dendritic cell. [9] IFN-γ and other cytokines released by activated leukocytes enhance local epithelial class II expression. [10] Such cytokines also induce terminal B-cell differentiation and [11] increased expression of secretory component, thereby [12] promoting external transport of dimeric IgA.

The preferential migration of T cells into mucosal epithelia has yet to be explained. Most intraepithelial lymphocytes (IEL) express a unique surface membrane molecule, but it does not seem to be involved in their migration into the gut epithelium (Cerf-Bensussan et al. 1988). This migration is partly antigen-independent because IEL are observed before birth (Spencer et al. 1986a); nevertheless, luminal antigens clearly determine the magnitude of migration (Ferguson and Parrott 1972). The follicle-associated epithelium (FAE) covering the domes of human Peyer's patches (PP) contains particularly many T cells, especially near the antigen-transporting M ("membrane") cells (Bjerke et al. 1988).

Nature of Intraepithelial Lymphocytes

IEL appearing in the gut epithelium are mostly $CD3^+$ and show a striking predominance of T8 ($CD8^+$) cells (cytotoxic/suppressor phenotype), whereas the lamina propria contains mainly the T4 ($CD4^+$) (helper/inducer) subset (Brandtzaeg et al. 1988). Bronchial and nasal mucosae contain fewer IEL than the gut, and T4 cells are remarkably predominant. Also the FAE of human PP contains a much higher proportion of T4 cells than the intestinal villous epithelium, i.e. ~40% *vs.* ~10% (Bjerke et al. 1988).

In the epithelium of the normal jejunum, a "blast marker" (CD7) is expressed mainly on the T8 cells, indicating that they are stimulated (Trejdosiewicz et al. 1987), which is in line with morphological features (Marsh 1975). However, most IEL seem to be negative for typical activation markers such as major histocompatibility complex (MHC) class II and Tac antigens. In both the jejunum and colon the IEL are also mainly negative for the H366 antigen of cytotoxic cells, suggesting that most of the intraepithelial T8 cells are functional suppressor (Ts) cells (Trejdosiewicz et al. 1987). As the IEL are often found along the basement membrane, apparently crossing it in either direction, they usually seem to leave the epithelium quite rapidly by reentering the lamina propria (Marsh 1975). It is possible that they mainly perform their putative immunoregulatory function(s) in the latter microenvironment.

IgA-PROMOTING CELLULAR INTERACTIONS

Clonal immaturity, as evidence by potential for J-chain production, combined with prominent IgA expression, characterizes the B cells that migrate to secretory tissues from GALT and BALT (Brandtzaeg 1985). It is not fully understood how these phenotypic features are favored by the mucosal microenvironments. The appropriate regulatory events probably take place both in organized lymphoepithelial structures such as PP, and additionally in the secretory tissues (Fig. 1). At the former sites some local mechanism(s) apparently exists that facilitates expansion of early memory clones while inducing little terminal B-cell differentiation (Bjerke and Brandtzaeg 1986).

T Cells

Most information is available for PP of experimental animals (reviewed in Brandtzaeg et al. 1987a). Helper T (Th) cells that mediate B-cell isotype switching directly from IgM to IgA expression have been cloned from murine PP (Strober and Sneller 1988). Observations on human B cells have indicated that this direct pathway may preferentially lead to IgA2 production (Conley and Bartelt 1984), which is the predominant isotype in the distal gut (Kett et al. 1986). The switch Th cells seem to be autoreactive and are apparently triggered directly by MHC class II determinants. *In vitro* experiments with human appendiceal lymphoid cells have supported an important role of class II (HLA-DR) molecules in the induction of IgA and IgM production (Kawanishi 1987). HLA-DR determinants are expressed both on FAE (Bjerke and Brandtzaeg 1988) and on villous epithelium in the normal human gut, and also on various histiocytic cells and B cells; it is therefore not possible as yet to know which accessory element(s) is most important for mucosal T-cell regulation *in vivo.*

The murine switch Th cells are apparently unable to induce terminal B-cell differentiation, but Fcα receptor-positive post-switch Th cells favoring IgA production have also been cloned from murine PP (Mestecky and McGhee 1987). Th cells that promote IgA production by both switch and post-switch effects have recently been cloned from the human appendix (Strober and Sneller 1988).

Lymphokines

The involvement of various T cell-derived lymphokines in the preferential IgA induction is currently in the focus of interest, although it is unknown at which regulatory level these peptides act (Harriman and Strober 1987). The *in vitro* IgA-promoting effect of interleukin 5 (IL-5) is particularly striking, both in the mouse and human. Synergistic interactions between various lymphokines are probably important, for example, in the induction of IL-2 receptors on B cells by IL-5 (O'Garra et al. 1988). In this context it is of interest that IL-2 and interferon-γ (IFN-γ) have been shown to synergize in selective induction of an IgA response to a bacterial polysaccharide (Murray et al. 1985). Activated T-cells may also induce transcription of the J-chain gene via IL-2 (Blackman et al. 1986).

Accessory Cells

Despite these fascinating observations, a comprehensive and generally accepted explanation of mucosal B-cell regulation is as yet not available. The importance of particular accessory cells within GALT have been emphasized in several studies (reviewed in Brandtzaeg et al. 1987a). Certain dendritic cells in combination with T lymphocytes from murine PP were thus found to remarkably augment IgA production in contrast to similar cells from spleen. It is possible that antigen-presenting cells (APC) in GALT and MALT preferentially attract or activate Th cells with a high level of IL-5 secretion. Such T cells are to some extent autoreactive and may thus be triggered by MHC class II-positive accessory cells.

The exact nature of the crucial accessory cells and their subepithelial, inter- or intra-follicular distribution within GALT and BALT are unknown (Fig. 1). Although it is uncertain to what extent germinal center blasts are sufficiently immunoresponsive and migratory for significant contribution to the population of IgA cells seeding secretory sites (Lebman et al. 1987), it is interesting that studies with monoclonal antibodies have indicated that follicular dendritic cells present in GALT and BALT differ phenotypically from those in spleen and lymph nodes; and those in GALT differ from those in BALT. Also, a certain macrophage population present in spleen and lymph nodes seems to be absent in GALT and BALT (Sminia et al. 1989).

POSITIVE TERMINAL IMMUNE REGULATION AT SECRETORY SITES

After extravasation in secretory tissues, the B cells are subjected to additional immune regulation, which induces both proliferation and isotype switching. Moreover, secondary signals for terminal differentiation into Ig-producing plasma cells are mainly provided at these sites (reviewed in Brandtzaeg et al. 1987a).

Putative Role of MHC Class II-Positive Epithelium

There is good evidence to suggest that the magnitude of immune responses is related to the density of MHC class II determinants (Janeway et al. 1984). It is therefore tempting to speculate that HLA-DR determinants not only on dendritic APC in the lamina propria, but also on epithelial cells contribute to topical B-cell differentiaion (Fig. 1). To this end epithelial DR-expression may be particularly important in secretory tissues remote from mucosae, such as mammary and major salivary glands where little or no foreign antigen gains access (Brandtzaeg et al. 1988). IgA-producing cells tend to accumulate near DR-positive ducts in parotid and submandibular glands, and the alveolar cells of lactating mammary glands are strongly DR-positive. Nevertheless, the number of IgA-producing cells per tissue unit is much greater in the intestinal mucosa and lacrimal glands which are near mucosal surfaces with massive antigen exposure (Fig. 2).

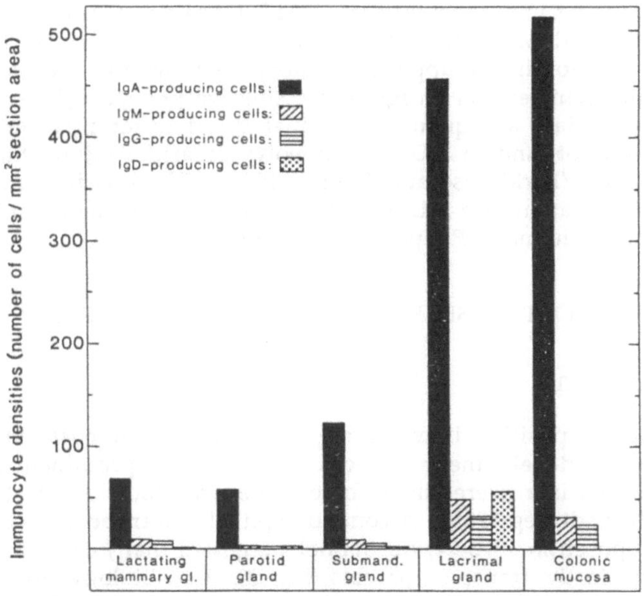

Fig. 2. Densities of Ig-producing immunocytes of different classes in various normal human secretory tissues. Schematic representation based on immunohistochemical data from the author's laboratory.

Disparity within the Secretory Immune System

As shown in Fig. 2, there is a striking difference between the lacrimal gland and the colonic mucosa with regard to the number of IgD-producing cells; the latter class of immunocytes is in fact virtually lacking in the gastrointestinal mucosa whereas they are substantially represented (2-10%) in the upper respiratory and digestive tracts, including nasal mucosa and lacrimal, salivary, and mammary glands, in that order (Brandtzaeg 1983).

Such disparity is even more striking in patients with selective IgA deficiency (Brandtzaeg et al. 1987b). These observations, along with a comparable incongruity noted for the local production of IgA subclasses (Kett et al. 1986), suggest that immunoregulatory mechanisms operating in GALT and BALT (including the tonsils) to some extent are different. We have proposed that the development of IgA immunocytes may largely depend on direct switching from IgM to IgA2 expression in the distal gut, whereas more sequential heavy-chain switching favoring IgA1 takes place in the upper respiratory mucosa and lacrimal, salivary and mammary glands - in decreasing order. J chain-positive IgD and IgG immunocytes, which are relatively frequent in the latter tissue sites (Brandtzaeg 1985), may be considered as clonal "spin-off" from a sequential B-cell differentiation process.

The notion that B cells of BALT and tonsils have appreciably more IgD-producing potential than those from GALT, is supported by the fact that tonsils normally contain far more IgD-producing immunocytes than GALT (Bjerke and Brandtzaeg 1986). This difference may depend on environmental rather than on intrinsic factors. Many bacterial species common in the upper respiratory tract, have Fc receptors for IgD and may thus exert mitogenic effects on IgD-expressing lymphocytes (Forsgren and Grubb 1979). This may initiate a sequential terminal B-cell differentiation. By contrast, *Escherichia coli* does not bind to IgD and lipopolysaccharides inhibit the expression of IgD on B cells *in vitro* (Parkhouse and Cooper 1977). There is immunohistochemical evidence suggesting that surface IgD is preferentially downregulated at a certain B-cell differentiation stage in human PP (Spencer et al. 1986b).

EPITHELIAL TRANSPORT OF SECRETORY IMMUNOGLOBULINS

SC as a Receptor Protein

The ultimate goal of positive immune regulation in the mucosal immune system, must be to favor the development of B-cell clones with a prominent potential for J-chain expression; this is a prerequisite for external translocation of locally produced pIgA and pIgM (Brandtzaeg 1985). A common epithelial transport model for these two polymers was proposed in 1973-74; our model suggested that J chain and SC represent the "lock and key" in the transport process (Fig. 3). Cell biology studies have shown that rabbit SC is produced as a transmembrane protein family 25-30 kD larger than secreted free SC. The relatively large cytoplasmic (~15 kD) portion seems to function as an effector domain guiding SC on its migration through the cell, as recently revealed in experiments with mutant transmembrane SC (Breitfeld et al. 1988).

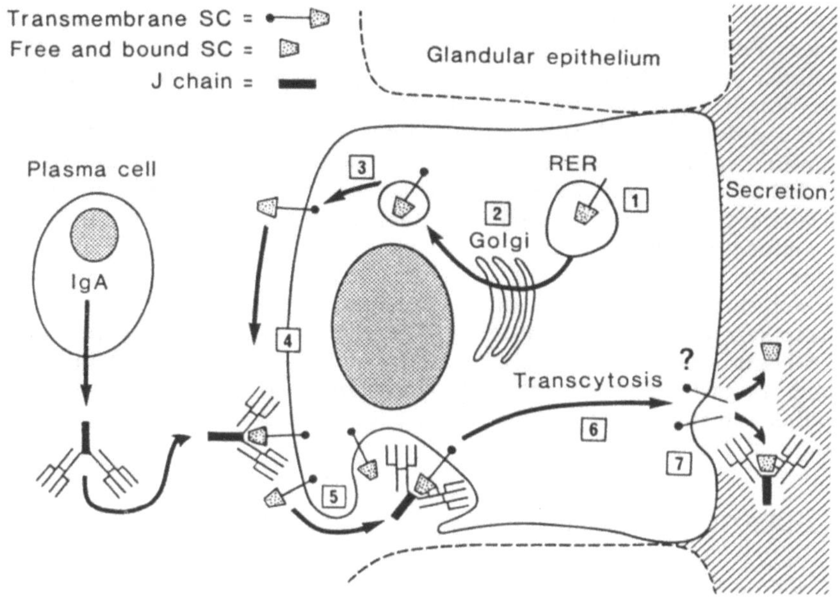

Fig. 3. Model for SC-mediated epithelial transport of dimeric IgA. [1] synthesis and core glycosylation of transmembrane SC in rough endoplasmic reticulum (*RER*) of epithelial cell. [2] Terminal glycosylation (●) in the Golgi complex. [3] Expression of SC at basolateral cell surfaces. [4] Complexing of SC with J chain-containing dimeric IgA. [5] Endocytosis of ligand-receptor complexes and excess SC. [6] Transcytosis of vesicles. [7] Cleavage and release of SIgA and free SC. The cleavage mechanism and the fate of the cytoplasmic tail of transmembrane SC are unknown (?). During the external translocation, covalent stabilization of the IgA-SC complexes regularly occurs (two disulfide bridges indicated in SIgA between SC and one of the IgA subunits).

The production of core glycosylated transmembrane SC takes place on the rough endoplasmic reticulum and needs 30-60 min for maturation in the Golgi apparatus (Fig. 3); the subsequent migration to the basolateral plasma membrane and the transport of pIgA to the luminal cell face apparently occur very quickly, and this route seems to be followed by at least 90% of the SC (Breitfeld et al. 1988). Experiments with the HT-29 colonic carcinoma cell line have demonstrated that human SC is likewise produced as a transmembrane precursor (~95 kD) which, by addition of sugars, becomes about 20 kD larger than the secreted form (Mostov and Blobel 1982). Also transmembrane SC depends on the presence of J chain in pIgA and pIgM to perform specific epithelial uptake of these polymers (Brandtzaeg 1985).

Immunoregulatory Aspects of SC-Mediated Transport

Cloned cDNA of rabbit SC showed that it contains five extra-cellular domains with remarkable homology to each other and to immunoglobulin variable or constant region domains (Breitfeld et al. 1988). The amino acid sequence of human free SC suggests the existence similar homology regions (Eiffert et al. 1984). SC thus belongs to the "Ig-superfamily" which includes a variety of cell-surface recognition structures.

Like other members of the Ig superfamily, SC expression can be regulated by lymphokines. Both recombinant IFN-γ and tumor necrosis factor-α (TFN-α) upregulate the intracellular pool and epithelial membrane expression of functional SC in the HT-29 cell line (Brandtzaeg et al. 1988). The effects of these two cytokines on SC expression are additive in a dose-dependent manner (Kvale et al. 1988). IFN-γ is secreted by T cells during immune responses whereas activated macrophages are the best recognized source of TFN-α. Our observations suggest that both these cell types may promote the external transport of pIgA and pIgM and thereby enhance the efferent limb of the secretory immune system (Fig. 1).

Our immunohistochemical studies on human gastric mucosa are consistent with an immune response-related upregulation of SC-mediated epithelial transport. Signs of increased SC expression and enhanced uptake of IgA were observed in fundic and antral glands in gastritis specimens with dense lymphoid infiltration; we have noted similar epithelial features also in celiac disease lesions and in inflamed salivary glands (Brandtzaeg et al. 1988).

DOWNREGULATION OF POTENTIALLY HARMFUL IMMUNE RESPONSES

Induction of Oral Tolerance

When the secretory immune system is unable to perform adequate antigen exclusion, the internal body environment should preferably be protected against potentially harmful systemic types of immune reactions elicited by IgG, IgE and T cell-mediated delayed-type hypersensitivity (DTH). There is experimental evidence in animals suggesting that such protection may be afforded by immunosuppressive mechanisms referred to as "oral tolerance" (Mowat 1987). This phenomenon probably involves multiple immunoregulatory events which in part may be different for humoral immunity and DTH. Antigen handling by an intact gut epithelium seems to be critical (Nicklin and Miller 1983), and a role of intestinal antigen "processing" has been suggested - a least for suppression of DTH (Mowat 1987). The nature of such processing and the cellular element(s) involved (? epithelium or special mucosal macrophages), are still obscure.

Putative Role of Lympho-Epithelial Interactions

In view of recent information about lympho-epithelial interactions (Brandtzaeg et al. 1988), it is tempting to propose that suppression mediated by intraepithelial T8 cells may be an important aspect of oral tolerance (Fig. 4). Bland and Warren (1986) found that MHC class II-positive columnar cells from rat intestinal villous epithelium could present ovalbumin to primed lymph node T cells which thereby were induced to proliferate; this apparently class II-restricted immune response led to antigen-specific suppression. The phenotype of the inducer cell was suggested to belong to a T8 subset. Mayer and Shlien (1987) applied human colonic epithelial cells in autologous or allogeneic mixed lymphocyte responses and found likewise preferential stimulation of T8 cells. These lymphocytes did not express the 9.3 antigen, which is a marker of cytotoxic T cells; and they showed no cytotoxic effect but performed, instead, non-specific suppression. It was possible to block the response with a rabbit antibody to human MHC class II determinants, indicating that such molecules in some way were involved in the induction phase.

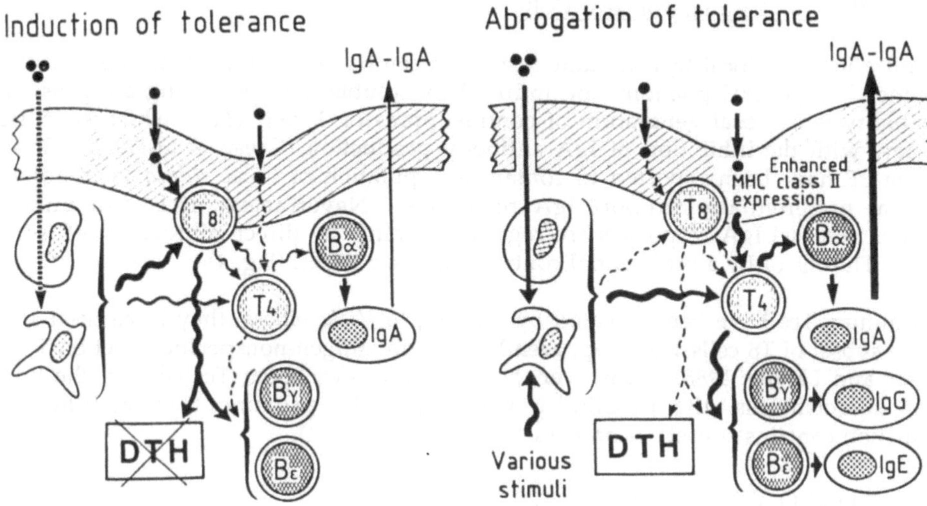

Fig. 4. Hypothetical scheme for induction and abrogation of locally induced tolerance which normally affords downregulation of both delayed type hypersensitivity (DTH) and IgG (Bγ) and IgE (Bε) responses with concomitant maintenance of IgA (Bα) response. *Left*: Presentation to T8 suppressor cells of antigen that has been processed by epithelial cell and/or by specialized mucosal macrophage, or direct epithelial antigen presentation, results in dominating regulatory signals conducive to suppression (*heavy waved split arrow*). Little or no help is provided by T4 cells to Bγ and Bε cells (*broken waved arrow*), whereas Bα cells receive substantial help from appropriate T4 cells (*waved thin arrow*); the latter cells are either preferentially stimulated by antigen-presenting dendritic and/or epithelial cells or released from T8-induced suppression by signals from a contrasuppressor T8 subset. *Right*: When there is a break in the epithelial lining, undue stimulation of antigen-presenting cells, and/or enhanced epithelial expression of MHC class II determinants with ensuing increased presentation of luminal antigens, the result is general overstimulation of T4 cells (*changed quality of arrows*). This may give rise not only to increased production of dimeric IgA, but also to excessive IgG (and IgE) responses and DTH.

Since T8 lymphocytes are the predominant IEL along the normal gut wall, the above observations suggest that these cells are responsible for epithelium-dependent suppression as part of a controlled mucosal immunoregulatory circuit (Fig. 4). However, much remains to be learned about interactions between gut epithelial cells and lymphocytes, both with regard to the requirement for putative third-party cells and the actual MHC restriction element(s) involved (Brandtzaeg et al. 1988). It cannot also be excluded that class II-reactive autostimulatory T4 cells are involved in a suppressive circuit (Pawelec et al. 1988).

Putative Role of Contrasuppressor Cells

It is puzzling how local IgA immune responses can be released from suppression if oral tolerance is a general phenomenon induced by soluble non-replicating antigens. One possiblility is mucosal generation of contrasuppressor T cells (Tcs), which specifically interfere with the inhibition of IgA antibody production (Green et al. 1988). Indeed, murine PP have been reported to contain a population of *Vicia villosa* lecin-adherent Tcs that preferentially support IgA responses. Nevertheless, it is still unknown whether mucosal IgA responses are regulated mainly by direct stimulation of isotype-specific Th cells or by a preferential contrasuppressor circuit (Fig. 4).

Also human Tcs have been isolated by binding to *Vicia villosa*; they have been shown to be a subset of T8 cells which apparently exert an antigen-nonspecific effect on Th cells (Brines and Lehner 1988). However, nothing is known about Tcs cells in the human gut. Nevertheless, the possibility exists that the prominent intraepithelial T8 population contains subsets of both Ts and Tcs cells (Fig. 4).

Abrogation of Oral Tolerance

General abrogation of tolerance to luminal antigens may be involved in the pathogenesis of a variety of mucosal diseases (Fig. 4), as suggested by local overproduction of IgG (Brandtzaeg et al. 1987a) and decreased J-chain expression (Kett et al. 1988). Moreover, a regional IgE response with an abundance of IgE-positive mucosal mast cells may be seen in subjects with type I (atopic) allergy (Brandtzaeg et al. 1987a). In experimental animals oral antigen feeding combined with some sort of damage to the gut epithelium (Nicklin and Miller 1983) - or direct injection of antigen into the PP (Dunkley and Husband 1987) - seems to be incompatible with induction of suppression. The same is true when the antigen-presenting cells are excessively activated by stimuli such as muramyl dipeptide, estrogen, or a graft-versus-host reaction (Mowat 1987). Both situations apparently favor general overstimulation of Th cells.

Also aberrant epithelial MHC class II expression may be involved in abrogation of tolerance. Preferential activation of T4 rather than T8 lymphocytes was observed after stimlulation with colonic epithelial cells from inflammatory bowel disease (Mayer and Eisenhardt 1987), in which epithelial HLA-DR is markedly increased. Enhanced and differential epithelial class II expression may likewise be involved in exaggerated immune responses to gluten and other dietary antigens in celiac disease and to autoantigens in Sjögren's syndrome (Brandtzaeg et al. 1988). The lymphoepithelial interactions may in this way contribute to class II-related predisposition apparently expressed in the latter diseases.

CONCLUSIONS

The efferent limb of secretory immunity depends on local production and selective epithelial transport of pIgA (and pIgM). To this end there is a fascinating co-operation between the B-cell system and the epithelium by their production of two key factors: J chain and SC, respectively. Local immune regulation must be geared for generation of early memory B-cell clones with prominent J-chain expression and preference for IgA production. Interactions between activated leukocytes (e.g., T cells and macrophages) and the secretory epithelium probably contribute to enhancement of the SC-dependent transport of J chain-containing pIg molecules. Evidence is accumulating, moreover, to suggest that lympho-epithelial interactions are also involved in T cell-mediated mucosal induction of tolerance to systemic type of immunity to luminal antigens. IgG (and IgE) responses and also delayed-type hypersensitivity, which all may produce undue inflammation and tissue damage, will thereby be minimized. There are probably, in addition, complex interactions between the mucosal immune system and neuroendocrine cells, which have not been considered in this review.

Acknowledgments. Supported by grants from the Norwegian Research Council for Science and the Humanities and the Norwegian Cancer Society.

REFERENCES

Besredka A (1919) De la vaccination contre les etats typoides par voie buccale. Ann Ist Pasteur 33:882-903

Bjerke K, Brandtzaeg P (1986) Immunoglobulin- and J chain-producing cells associated with lymphoid follicles in the human appendix, colon and ileum, including Peyer's patches. Clin Exp Immunol 64:432-441

Bjerke K, Brandtzaeg P (1988) Lack of relation between expression of HLA-DR and secretory component (SC) in follicle-associated epithelium of human Peyer's patches. Clin Exp Immunol 71:502-507

Bjerke K, Brandtzaeg P, Fausa O (In press) T-cell distribution is different in follicle-associated epithelium of human Peyer's patches and villous epithelium. Clin Exp Immunol

Blackman MA, Tiggs MA, Minie ME, Koshland ME (1986) A model system for peptide-hormone action in differentiation. Interleukin-2 induces a B lymphoma to transcribe the J chain gene. Cell 47:609-617

Bland PW, Warren LG (1986) Antigen presentation by epithelial cells of the rat small intestine. II. Selective induction of suppressor T cells. Immunology 58:9-14

Brandtzaeg P (1983) The secretory immune system of lactating human mammary glands compared with other exocrine organs. Ann N.Y. Acad Sci 409:353-381

Brandtzaeg P (1985) Role of J chain and secretory component in receptor-mediated glandular and hepatic transport of immunoglobulins in man. Scand J Immunol 22:111-146

Brandtzaeg P, Baklien K, Rognum TO, Scott H, Valnes K (1987a) Nature and properties of the human gastrointestinal immune system. In: Miller K, Nicklin S (eds) Immunology of the gastrointestinal tract, Vol. 1, CRC Press, Boca Raton, Florida, p 1

Brandtzaeg P, Karlsson G, Hansson G, Petruson B, Björkander J, Hanson LÅ (1987b) The clinical condition of IgA-deficient patients is related to the proportion of IgD- and IgM-producing cells in their nasal mucosa. Clin Exp Immunol 67:626-636

Brandtzaeg P, Sollid LM, Thrane PS, Kvale D, Bjerke K, Scott H, Kett K, Rognum TO (1988) Lymphoepithelial interactions in the mucosal immune system. Gut 29:1116-1130

Breitfeld PP, Harris JM, Mostov KE (1988) Cell biology of the IgA receptor in polarized epithelia. Monogr. Allergy 24:66-70

Brines R, Lehner T (1988) Characterization of human T8$^+$ suppressor and contrasuppressor cells, separated by the lectin *Vicia villosa*. Immunology 63:247-254

Cerf-Bensussan N, Jarry A, Gnéragbé T, Brousse N. Lisowska-Grospierre B, Griscelli C, Guy-Grand D (1988) Monoclonal antibodies specific for intestinal lymphocytes. Monogr Allergy 24:167-172

Conley ME, Bartelt MS (1984) *In vitro* regulation of IgA subclass systhesis. II. The source of IgA2 plasma cells. J Immunol 133:2311-2316

Davies A (1922) An investigation into the serological properties of dysentery stools. Lancet ii:1009-1012

Delacroix DL (1985) The human immunoglobulin A system: Its vascular compartment. Thesis, Universite Catholique de Louvain, European Medical Press, Brugge, Belgium

Dunkley ML, Husband AJ (1987) Distribution and functional characteristics of antigen-specific helper T cells arising after Peyer's patch immunization. Immunology 61:475-482

Eiffert H. Quentin E, Decker J, Hillemeir S, Hufschmidt M, Klingmüller D, Weber MH, Hilschmann N (1984) Die Primärstruktur der menschlichen freien Secretkomponente und die Anordnung der Disulfidbrücken. Hoppe-Seyler's Z Physiol Chem 365:1489-1495

Ferguson A, Parrott DMV (1972) The effect of antigen deprivation on thymus-dependent and thymus-independent lymphocytes in the small intestine of the mouse. Clin Exp Immunol 12:477-488

Forsgren A, Grubb AO (1979) Many bacterial species bind human IgG. J Immunol 122:1468-1472

Green DR, McGhee J, Ptak W (1988) Contrasuppression and the control of microenvironmental immune responses. Monogr Allergy 24:236-243

Harriman GR, Strober W (1987) Interleukin 5, a mucosal lymphokine ? J Immunol 139:3553-3555

Janeway CA, Bottomly K, Babich J, Conrad P, Conzen S, Jones B, Kaye J, Katz M, McVay L, Murphy DB, Tite J (1984) Quantitative variations in Ia antigen expression plays a central role in immune regulation. Immunol Today 5:99-105

Kawanishi H (1987) Immunocompentence of normal human appendiceal lymphoid cells: *In vitro* studies. Immunology 60:19-28

Kett K, Brandtzaeg P, Radl J, Haaijman JF (1986) Different subclass distribution of IgA-producing cells in human lymphoid organs and various secretory tissues. J Immunol 136:3631-3635

Kett K, Brandtzaeg P, Fausa O (1988) J-chain expression is more prominent in immunoglubulin A2 than in immunoglobulin A1 colonic immunocytes and is decreased in both subclasses associated with inflammatory bowel disease. Gastroenterology 94:1419-1425

Kvale D, Brandtzaeg P, Lövhaug D (In Press) Tumor necreosis factor-α and interferon-γ upregulate co-operatively but differently the expression of secretory component and HLA molecules in a human colonic cell line. Scand J Immunol

Lebman DA, Griffin PM, Cebra JJ (1987) Relationship between expression of IgA by Peyer's patch cells and functional IgA memory cells. J Exp Med 166:1405-1418

Marsh MN (1975) Studies of intestinal lymphoid tissue. II. Aspects of proliferation and migration of epithelial lymphocytes in the small intestine of mice. Gut 16:674-682

Mayer L, Eisenhardt D (1987) Lack of induction of supressor T cells by gut epithelial cells from patients with inflammatory bowel disease. The primary defect? Gastroenterology 92:1524 (Abstract)

Mayer L, Shlien R (1987) Evidence for function of Ia molecules on gut epithelial cells in man. J Exp Med 166:1471-1483

Mestecky J, McGhee JR (1987) Immunoglobulin A (IgA): Molecular and cellular interactions involved in IgA biosynthesis and immune response. Adv Immunol 40:153-245

Mostov KE, Blobel G (1982) A transmembrane precursor of secretory component. The receptor for transcellular transport of polymeric immunoglobulins. J Biol Chem 257:11816-11821

Mowat AM (1987) The regulation of immune responses to dietary protein antigens. Immunol Today 8:93-98

Murray PD, Swain SL, Kagnoff MF (1985) Regulation of the IgM and IgA anti-dextran B1355S response: Synergy between IFN-γ, BCGF II, and IL 2. J Immunol 135:4015-4020

Nicklin S, Miller K (1983) Local and systemic immune responses to intestinally presented antigen. Int Archs Allergy Appl Immun 72:87-90

O'Garra A, Umland S, De France T, Christiansen J (1988) 'B-cell factors' are pleiotropic. Immunol Today 9:45-54

Parkhouse RME, Cooper MD (1977) A model for the differentiation of B lymphcytes with implications for the biological role of IgD. Immunol Rev 37:105-126

Pawelec G, Fernandez N, Brocker T, Schneider EM, Festenstein H, Wernet P (1988) DY determinants, possibly associated with novel class II molecules, stimulate autoreactive $CD4^+$ T cells with suppressive activity. J Exp Med 167:243-261

Sminia T, van der Brugge-Gammelkoorn GJ, Jeurissen SHM (1989) Structure and function of bronchus-associated lymphoid tissue (BALT). CRC Critical Rev Immunol 9 (Issue 2)

Spencer J, Dillon SB, Isaacson PG, MacDonald TT (1986a) T cell subclasses in fetal human ileum. Clin Exp Immunol 65:553-558

Spencer J, Finn T, Isaacson PG (1986b) Human Peyer's patches: An immunohistochemical study. Gut 27:405-410

Streeter PR, Berg EL, Rouse BTN, Bargatze RF, Butcher EC (1988) A tissue-specific endothelial cell molecule involved in lymphocyte homing. Nature 331:41-46

Strober W, Sneller MC (1988) Cellular and molecular events accompanying IgA B cell differentiaiton. Monogr Allergy 24:181-190

Tomasi TB, Tan EM, Solomon A, Prendergast RA (1965) Characteristics of an immune system common to certain external secretions. J Exp Med 121:101-124

Trejdosiewicz LK, Malizia G, Badr-el-Din S, Smart CJ, Oakes DJ, Southgate J, Howdle PD, Janossy G (1987) T cell and mononuclear phagocyte populations of the human small and large intestine. Adv Exp Med Biol 216A:465-473

van der Heijden PJ, Stok W, Bianchi TJ (1987) Contribution of immunoglobulin-secreting cells in the murine small intestine to the 'background' immunoglobulin production. Immunology 62:551-555

Antigen Delivery System

Cholera Toxin and its Subunits as Potential Oral Adjuvants

C.O. Elson

Division of Gastroenterology, University of Alabama at Birmingham, Birmingham, AL 35294, USA

INTRODUCTION

Cholera toxin (CT) is a protein produced by *Vibrio cholerae* whose effect on the intestinal epithelial cells is responsible for the copious fluid secretion seen in clinical cholera. Because of this, the biology, immunology, and biochemistry of this molecule have been extensively characterized (Holmgren 1981; van Heyningen 1982). The molecular structure of CT is known and the genes encoding it have been isolated and sequenced (Mekalanos et al. 1983; Lockman and Kaper 1983). CT has two subunits, a 28,000 kD A subunit that has ADP-ribosylating enzymatic activity and an 11,500 kD B or binding subunit which exists as a pentamer in the holotoxin. The receptor molecule for CT is known to be GM1 ganglioside, a constituent of the cell surface membrane of virtually all nucleated cells. The A subunit can be further subdivided into A1 and A2 components which are encoded in the same gene segment and are linked by a disulfide bond. It is the hydrophobic A1 subunit that appears to insert into the membrane and mediate the activation of the adenylate cyclase enzyme.

PROPERTIES OF CHOLERA TOXIN AS AN ORAL IMMUNOGEN

Natural infection with *V. cholerae* results in an antibody response both to the vibrio and to CT itself (Holmgren 1981). Because antibodies can neutralize the effects of the toxin (Peterson et al. 1979), the immune response to the toxin has been studied extensively in experimental animals (Pierce and Gowans 1975; Pierce 1978; Lange et al. 1980; Holmgren 1981; Elson and Ealding 1984a, 1984b). Indeed, CT is the most potent non-viable oral immunogen yet identified (Pierce and Gowans 1975), and achieving a similar response with other types of antigens presented orally seems a worthwhile but as yet unattained goal. Some of the major properties of CT as an oral immunogen are presented in Table 1.

Table 1. Properties of CT as an oral immunogen

- Stimulates specific secretory IgA and plasma IgG
- Does not induce oral tolerance
- Response is I-A restricted in mice
- Regionalization of response
- Extended memory

Current Topics in Microbiology and Immunology, Vol. 146

Oral immunization with this protein results not only in a specific secretory IgA response but also in a plasma IgG response, both of which appear to originate in gut-associated-lymphoid tissue or GALT (Elson and Ealding 1984a). In contrast to other protein antigens, CT feeding does not induce oral tolerance (Elson and Ealding 1984b), which is certainly a desirable property for an oral immunogen. Both the secretory IgA and systemic IgG antibody responses to CT are restricted by the I-A subregion of the H-2 major histocompatibility locus (Elson and Ealding 1985, 1987). There is evidence for regionalization of the response, i.e., plasma cells producing anti-CT antibody are most abundant in areas of mucosa that are directly exposed to the toxin (Lange et al. 1980; Pierce and Cray 1982). This may represent an important consideration in oral immunization schemes where the major IgA response is desired in non-intestinal mucosal sites. Lastly, CT is able to induce a prolonged memory response at mucosal surfaces (Lycke and Holmgren 1987).

PROPERTIES OF CHOLERA TOXIN AS AN ORAL ADJUVANT

In contrast to the ease of oral immunization with CT, attempts to induce oral immunization with other protein antigens has been difficult, frustrating, and largely unsuccessful. For example, the feeding of mice with other protein antigens such as keyhole limpet hemocyanin (KLH), a strong immunogen when given parenterally, not only fails to stimulate secretory IgA responses, but instead induces oral tolerance to it (Elson and Ealding 1984a, 1984b). It has been of some interest therefore that the co-administration into the intestine of both KLH and CT results in a secretory IgA response to KLH and at the same time an abrogation of oral tolerance (Elson and Ealding 1984a, 1984b). The timing and route of antigen administration is important in this adjuvant property of CT in that adjuvanticity only occurs when both CT and antigen are given by the same route and at the same time (Lycke and Holmgren 1986). This oral adjuvant effect of CT or its subunits has now been demonstrated not only with KLH, but also with horseradish peroxidase (McKenzie and Halsey 1984), Sendai virus (Nedrud et al. 1987) and influenza virus (Chen this volume). All of this work has been done in mice, but because the response to CT as an immunogen is similar in both mouse and humans (Holmgren et al. 1977; Svennerholm et al. 1982; Svennerholm et al. 1984), it is anticipated that the similar oral adjuvanticity would occur also in humans and in other animals.

ROLE OF THE SUBUNITS IN THE PROPERTIES OF CHOLERA TOXIN

The relative importance of the two subunits in the immunogenicity of CT is not clear. The B subunit shares many of the properties of the holotoxin, including the ability to induce secretory IgA responses and the lack of induction of oral tolerance after feeding (Elson and Ealding 1984a). However, a consistent finding has been that the immune response to the holotoxin is much better than that to an equal amount of B subunit (Pierce 1978) which clearly suggests a role for the A subunit in the toxin's immunogenic properties. If the A subunit is a major element in the toxin's immunogenicity, it does not follow that this effect must be due to the A subunit's ability to activate adenylate cyclase, because a heat aggregated form of CT ("procholeragenoid") has very little toxicity but is a potent oral immunogen (Pierce et al. 1983).

The role of the subunits in the adjuvant properties of CT is also largely unknown, but again it appears that the holotoxin is much more effective than is the B subunit. In work done by McKenzie and Halsey (1984), intestinal administration of mixtures of CT B subunit plus horseradish peroxidase were ineffective in stimulating antibody responses to the latter in the intestine or in serum, whereas CT B subunit did have adjuvant effects when directly conjugated to horseradish peroxidase. Consistent with these data Lycke and Holmgren (1987) have found that the B subunit was unable to act as an adjuvant when mixed with KLH whereas the holotoxin was effective. Together these experiments indicate that the toxin is more effective than the B subunit, which implies that the A subunit plays a role in the toxin's adjuvanticity, but again this needs to be demonstrated experimentally.

WHAT IS THE MECHANISMS OF THE ADJUVANT EFFECT?

This is an intriguing question whose answer would allow the fashioning of new and rational strategies for oral immunization. The effects of CT when given parenterally with other antigens are complex; it can inhibit or stimulate various immune responses (Northrup and Fauci 1972; Chisari et al. 1974; Henney et al. 1973). Holmgren et al. (1974) showed some years ago that the addition of CT to lymphocytes *in vitro* was quite inhibitory. More recently, we have shown that the purified B subunit is inhibitory also for lymphocyte activation *in vitro*, although it is not as potent as CT (Woogen et al. 1987). We believe that at least part of the adjuvant properties of CT may be explained by its ability to inhibit suppressor cells in GALT. Consistent with this idea, we have recently shown in an adoptive transfer system that the feeding of KLH to mice generates suppressor T cells that inhibit both a secretory IgA and a plasma IgG responses to KLH. However, the feeding of both KLH and CT together to mice eliminates this suppression (Elson et al., unpublished data). There is also some earlier work that provides evidence for CT's ability to inhibit suppressor cells in a graft-versus-host system (Lange et al. 1978). However, the inhibition of suppressor cells would seem inadequate by itself to explain the strong positive responses that are seen when CT is co-administered with an antigen such as KLH. We have examined the adjuvant effect of CT using strains of mice that are either high or low responders for CT, measuring the response to KLH after the feeding of KLH alone versus after the feeding of KLH plus CT. Interestingly, intestinal sIgA anti-KLH responses in the group fed KLH plus CT occurred only in mice that are high responders to CT. Moreover, the intestinal sIgA anti-KLH response correlated highly with the sIgA anti-CT response in the same mice (Elson 1987). The occurrence of the adjuvant effect mainly in high responders for CT would seem to suggest that a strong helper response to CT in GALT was providing help for other antigens - e.g. KLH, that happened to be present at the same time; perhaps this is a form of "bystander help". The combination of suppressor cell-inhibition and "bystander help" is currently thought to underlay the adjuvant effects of the toxin.

PRODUCTION OF RECOMBINANT CHOLERA TOXIN CHIMERIC PROTEINS

Assuming that CT and its B subunit are effective oral adjuvants, the coupling of the genes for the toxin to antigens of interest might provide an effective strategy for the production of chimeric protein antigens in which each molecule contains both antigen and adjuvant. The feasibility of such an approach has recently been demonstrated by Dertzbaugh and Macrina (1987). These workers have constructed a set of cloning vectors using three different oligonucleotide linkers inserted upstream of a promoterless B subunit gene. Three linkers were developed in order to allow any gene to be inserted upstream and be in the correct reading frame with the B subunit gene.

Using these linkers a chimeric protein was produced containing a portion of the glucosyltransferase B enzyme produced by *Streptococcus mutans*. The chimeric protein bound to GM1 ganglioside and reacted on Western blots with antisera to both glucosyltransferase B and to B subunit. The efficacy of the chimera as an oral immunogen is not yet known. Similar manipulations could be performed using the A subunit. This approach represents a potential technology that could be used in the future to produce large amounts of effective oral vaccines.

SUMMARY

Cholera toxin has been shown to have adjuvant effects in multiple different systems. The dose, timing and genetic background of the recipient all seem to be important variables. The role of the two subunits in both the immunogenicity and the adjuvanticity of this molecule remain unclear. The mechanisms of the adjuvant effect likely involves effects on regulatory T cells; there is evidence that the adjuvant effect is due at least in part to inhibition of suppressor T cells. When KLH is used as a model antigen, the adjuvanticity of cholera toxin appears to be related to its immunogenicity in that both properties occur mainly in mouse strains that are high responders to cholera toxin. The genetic engineering of chimeric neoantigens consisting of cholera toxin subunits coupled to antigens of interest has been shown to be technically possible and is an attractive future approach for the generation of effective oral vaccines.

REFERENCES

Chisari FV, Northrup, Chen LC (1974) The modulating effect of cholera enterotoxin on the immune response. J Immunol 113:729-733

Dertzbaugh MT, Macrina FL (1987) Construction of novel cloning vectors for creating translational fusions with the B subunit of cholera toxin. Plasmid 18:256A

Elson CO (1987) Cholera toxin (CT) as a mucosal adjuvant - the effect of H-2 genes. Fed Proc 46:621A

Elson CO, Ealding W (1984a) Generalized systemic and mucosal immunity in mice after mucosal stimulation with cholera toxin. J Immunol 132:2736-2897

Elson CO, Ealding W (1984b) Cholera toxin feeding did not induce oral tolerance in mice and abrogated oral tolerance to an unrelated protein antigen. J Immunol 133:2892-2897

Elson CO, Ealding W (1985) Genetic control of the murine immune response to cholera toxin. J Immunol 135:930-932

Elson CO, Ealding W (1987) Ir gene control of the murine secretory IgA response to cholera toxin. Eur J Immunol 17:425-428

Henney CS, Lichtenstein LM, Gillespie E, Rolley RT (1973) *In vivo* suppression of the immune respone to alloantigen by cholera enterotoxin. J Clin Invest 52:2853-2864

Holmgren J (1981) Actions of cholera toxin and the prevention and treatment of cholera. Nature 292:413-417

Holmgren J, Lindholm L, Lonnroth I (1974) Interaction of cholera toxin and toxin derivatives with lymphocytes. I. Binding properties and interference with lectin-induced cellular stimulation. J Exp Med 139:801-809

Holmgren J, Svennerholm A-M, Lonnroth I, Fall-Persson M, Markman B, Lundbeck H (1977) Development of improved cholera vaccine based on subunit toxoid. Nature 269:602-604

Lange S, Lindholm L, Holmgren J (1978) Interaction of cholera toxin and toxin derivatives with lymphocytes. III. Modulating effects *in vivo* by cholera toxin on the graft-vs-host reactivity of lymphoid cells: Suggested inhibition of suppressor cells. Int Arch Allergy Appl Immunol 57:364-374

Lange S, Nygren H, Svennerholm A-M, Holmgren J (1980) Antitoxin cholera immunity in mice: Influence of antigen deposition on antitoxin-containing cells and protective immunity in different parts of the intestine. Infect Immun 28:17-23

Lockman H, Kaper JB (1983) Nucleotide sequence analysis of the A2 and B subunits of *Vibrio cholerae* enterotoxin. J Biol Chem 258:13722-13726

Lycke N, Holmgren J (1986) Strong adjuvant properties of cholera toxin on gut mucosal immune responses to orally presented antigens. Immunology 59:301-308

Lycke N, Holmgren J (1987) Long-term cholera antitoxin memory in the gut can be triggered to antibody formation associated with protection within hours of an oral challenge immunization. Scand J Immunol 25:407-412

McKenzie SJ, Halsey JF (1984) Cholera toxin B subunit as a carrier protein to stimulate a mucosal immune response. J Immunol 133:1818-1824

Mekalanos JJ, Swartz DJ, Pearson GDN, Harford N, Groyne F, de Wilde M (1983) Cholera toxin genes: Nucleotide sequence, deletion analysis and vaccine development. Nature 306:551-557

Mekalanos JJ, Swartz DJ, Pearson GDN, Harford N, Groyne F, de Wilde M (1983) Cholera toxin genes: Nucleotide sequence, deletion analysis and vaccine development. Nature 306:551-557

Nedrud JG, Mazanec MB, Liang X, Hague N, Lamm ME (1987) Induction and expression of respiratory IgA immunity against Sendai virus in mice. Adv Exp Med Biol 216B:1847-1854

Northrup RS, Fauci AS (1972) Adjuvant effect of cholera enterotoxin on the immune response of the mouse to sheep red blood cells. J Infect Dis 125:672-673

Peterson JW, Hejtmancik KE, Markel DE, Craig JP, Kurosky A (1979) Antigenic specificity of neutralizing antibody to cholera toxin. Infect Immun 24:774-779

Pierce NF and Cray WC (1982) Determinants of the localization, magnitude, and duration of a specific mucosal IgA plasma cell response in enterically immunized rats. J Immunol 128:1311-1315

Pierce NF, Gowans JL (1975) Cellular kinetics of the intestinal immune response to cholera toxoid in rats. J Exp Med 142:1550-1563

Pierce NF, Cray Jr WC, Sacci Jr JB, Craig JP, Germanier R, Furer E (1983) Procholeragenoid: A safe and effective antigen for oral immunization against experimental cholera. Infect Immun 40:1112-11118

Pierce NF (1978) The role of antigen form and function in the primary and secondary intestinal immune response to cholera toxin and toxoid in rats. J Exp Med 148:195-206

Svennerholm A-M, Jertborn M, Gothefors L, Karim AMMM, Sack DA, Holmgren J (1984) Mucosal antitoxic and antibacterial immunity after cholera disease and after immunization with a combined B subunit-whole cell vaccine. J Infect Dis 149:884-893

van Heyningen S (1982) Cholera Toxin. Bioscience Reports 2:135-146

Woogen SD, Ealding W and Elson CO (1987) Inhibition of murine lymphocyte proliferation by the B subunit of cholera toxin. J Immunol 139:3764-3770

Selective Delivery of Antigens by Recombinant Bacteria

R. Curtiss III, S.M. Kelly, P.A. Gulig, and K. Nakayama

Department of Biology, Washington University, St. Louis, MO 63130, USA

INTRODUCTION

All infectious disease agents have specific mechanisms to colonize, invade, and overcome a host. If this mechanism can be defined in biochemical terms, it should be possible to develop a vaccine to prevent infection and disease. Furthermore, understanding the means by which certain bacterial pathogens target to specific lymphoid tissues in an animal host enables one to develop novel strategies for targeting foreign antigens to that specific lymphoid tissue to elicit an immune response.

In the 1970's, it was learned that the delivery of antigens to the gut-associated lymphoid tissue (GALT or Peyer's patches) or to the bronchial-associated lymphoid tissue (BALT) leads to a generalized secretory immune response with production of secretory IgA (sIgA) (Cebra et al. 1976; Bienenstock et al. 1978; Weisz-Carrington et al. 1979; McCaughan and Basten 1983; LeFever and Joel 1984) in the secretions of the mammary, lachrymal, and salivary glands and fluids bathing all mucosal surfaces of the respiratory, gastrointestinal, and genito-urinary tracts (see Brandtzaeg this volume). Antigen presentation to the GALT is also able to elicit humoral (Asherson et al. 1977; Elson and Ealding 1984; Stevenson and Manning 1985; Dougan et al. 1986, 1987a, 1987b; Maskell et al. 1986, 1987; Black et al. 1987; Curtiss et al. 1987) and leads to cellular (Brown et al. 1987) immune responses.

It has been known for many years that *Salmonella typhi* in primates (Levy and Gaehtgens 1908; Muller 1912; Gaines et al. 1968) and *Salmonella typhimurium* in mice (Collins 1972; Carter and Collins 1974) are able to colonize the intestinal tract and to persist and proliferate in the GALT as well as in the liver and spleen prior to establishing a fatal bacteremia. In 1974, Carter and Collins conducted a careful study of the means by which *S. typhimurium* reached deep tissues in the mouse following ingestion. They conclusively demonstrated that *S. typhimurium* attached to, invaded, and persisted in the GALT as the first site of infection prior to reaching deep tissues. This led us (Curtiss et al. 1983, 1986; Curtiss 1985, 1986) and others (Dougan et al. 1986; Maskell et al. 1986) to consider the possibility of using attenuated *Salmonella* mutants as vectors to deliver specific antigens to the GALT.

In the early 1950's Bacon et al., demonstrated that *S. typhi* could be attenuated by imposing certain mutational defects resulting in auxotrophy for a number of essential metabolites. Subsequently, additional means to attenuate *Salmonella* were discovered (Germanier and Furer 1971, 1975; Hoiseth and Stocker 1981; Curtiss et al. 1987; Curtiss and Kelly 1987; McFarland and Stocker 1987), and it was shown that these avirulent *Salmonella* strains could, upon oral or intraperitoneal (i.p.) inoculation, induce immunity to subsequent challenge with fully virulent *Salmonella* strains (Germanier and Furer 1971, 1975; Hoiseth and Stocker 1981; Lindberg and Robertsson 1983; Robertsson et al. 1983; Killar and Eisenstein 1985; Mukkur et al. 1987; Hone et al. 1987).

Current Topics in Microbiology and Immunology, Vol. 146

With the advent of recombinant DNA techniques (Cohen et al. 1973) progress began to be made in elucidating specific mechanisms by which pathogens colonize on or invade through mucosal surfaces (Macrina 1984; Goebel 1985). Thus, one could consider having such foreign colonization or virulence antigens expressed by recombinant avirulent *Salmonella* and using these recombinant strains to target specific antigens to the GALT (Stevenson and Manning 1985; Dougan et al. 1986, 1987a; Maskell et al. 1986; Curtiss 1986; Curtiss et al. 1987).

AVIRULENT *SALMONELLA* MUTANTS

Plasmidless Mutants

Plasmids present in *S. typhimurium* (Jones et al. 1982; Hackett et al. 1986; Pardon et al. 1986; Gulig and Curtiss 1987), *S. dublin* (Terakado et al. 1983; Chikami et al. 1985; Manning et al. 1986), *S. enteritidis* (Nakamura et al. 1985a), and *S. gallinarum* (Barrow et al. 1987) contribute to the virulence of those pathogens. Although curing strains of these plasmids is difficult and can occur at exceedingly low frequencies (ca. 10^{-6} to 10^{-8} per bacterium per generation) (Gulig and Curtiss 1987) plasmidless derivatives have been isolated and found to have substantially reduced virulence when administered by the oral route. Nakamura et al. (1985b) were the first to suggest that such plasmidless strains might serve as effective attenuated vaccines against *Salmonella* infection. More recent work has demonstrated that plasmidless strains of *S. typhimurium* and *S. dublin* are capable of attaching to, invading, and persisting in the GALT while being defective in either reaching or surviving in the liver and spleen (Pardon et al. 1986; Hackett et al. 1986; Gulig and Curtiss 1987; Heffernan et al. 1987). Although there is some disagreement about the virulence attributes encoded by the plasmids in various *Salmonella* strains, progress is being made in defining the biochemical basis for plasmid-specified virulence by use of gene cloning and transposon mutagenesis (Manning et al. 1986; Hackett et al. 1987; Michiels et al. 1987; Gulig and Curtiss 1988).

In spite of the avirulence of plasmidless *Salmonella* strains when administered by the oral route, mice fed more than one hundred times the wild-type LD_{50} dose get sick, and some animals fed higher doses die (Gulig and Curtiss 1987). Because of this and the fact that plasmidless strains have essentially a wild-type LD_{50} when administered by the i.p. route (Gulig and Curtiss 1987), eliminating the virulence plasmid would be an insufficient means to develop a safe, efficacious vaccine strain.

Auxotrophic Mutants

Table 1 lists mutations rendering *Salmonella* avirulent without preventing immunogenicity. The mutants unable to synthesize *p*-aminobenzoic acid (*p*ABA) and purines were demonstrated to be avirulent by Bacon et al. (1950, 1951). Reversion to prototrophy was a potential safety problem with mutagen-induced auxotrophs. Thus, Hoiseth and Stocker (1981) made a significant contribution by introducing the use of transposon mutagenesis as a means to generate mutants with stable mutations blocking the aromatic pathway. The *aroA* mutants isolated by them had requirements for aromatic amino acids, which could be supplied by an animal host, and *p*ABA and dihydroxybenzoic acid, which could not; it is these last two requirements that supposedly rendered *aroA Salmonella* avirulent. The transposon Tn*10*, which confers resistance to tetracycline (Kleckner et al. 1977), affords the opportunity to selectively recover variants in which the Tn*10* and adjacent sequences have been lost to result in a deletion (Δ) mutation. Such low frequency excision events can be selected with fusaric acid, since fusaric acid resistance is associated with loss of tetracycline resistance (Bochner et al. 1980; Maloy and Nunn 1981). The use of transposon mutagenesis also provides a genetic marker (i.e., antibiotic resistance) to aid in moving the mutant allele from one strain to another by transduction (Schmieger 1972) or conjugation (Sanderson et al. 1972). Since *Escherichia coli* strains with requirements for diaminopimelic acid (DAP) and thymidine exhibited considerable biological containment and were deemed safe for use in potentially hazardous recombinant DNA experiments some 15 years ago (Curtiss et al. 1976, 1977; Curtiss 1978), we decided to investigate whether *Salmonella* mutants with requirements for DAP and/or thymidine were similarly avirulent yet immunogenic (Curtiss 1986; Curtiss et al. 1986, 1987).

Table 1. Mutations rendering *Salmonella* avirulent

Gene	Mutant phenotype	Reference
pab	Requirement for *p*ABA	Bacon et al. (1950, 1951)
pur	Requirement for purines	Bacon et al. (1950, 1951)
galE	Renders cells reversibly rough	Germanier and Furer (1971)
aroA	Requirement for aromatic amino acids, *p*ABA and dihydroxybenzoic acid	Hoiseth and Stocker (1981)
asd	Requirement for threonine, methionine, and diaminopimelic acid	Curtiss et al. (1986)
Ts	Decrease cell proliferation at 37° C	Ohta et al. (1987)
cya *crp*	Inefficient transport and use of carbohydrates and amino acids and inability to synthesize cell surface structures	Curtiss and Kelly (1987)

Δ*aroA* mutants of *S. typhimurium* (Hoiseth and Stocker 1981; Lindberg and Robertsson 1983; Robertsson et al. 1983; Habasha et al. 1985; Killar and Eisenstein 1985; Mukkur et al. 1987) and *S. dublin* (Smith et al. 1984), are avirulent and immunogenic to mice and other animal species. Purine-requiring mutants are also avirulent but may be too attenuated to induce an adequate protective immune response (McFarland and Stocker 1987). Δ*asd* mutants are totally avirulent by all routes of inoculation but survive for such a short period of time in the GALT of inoculated mice that they are only capable of inducing a secretory immune response and little or no humoral or cellular immune response (Curtiss et al. 1987). Δ*thyA* mutants are only mildly attenuated and therefore have not been tested for immunogenicity (Curtiss et al. 1987). A difficulty noted by Bacon et al. (1951) with auxotrophs is the potential to phenotypically reverse the attenuating mutation by inclusion of the required nutrient in the diet of animals inoculated with the auxotropic mutant. For this reason, strains with two or more mutations have been considered, but results with *S. typhi* Δ*aroA* Δ*purA* mutants used to immunize humans have been disappointing because of the severe attenuation imposed by this combination of mutations (Levine et al. 1987).

Mutants Altered in the Utilization and Synthesis of Carbohydrates

Germanier and Furer demonstrated that *galE* mutants of *S. typhimurium* (1971) and *S. typhi* (1975), which are unable to synthesize the enzyme UDP-galactose epimerase, were avirulent and immunogenic. The mechanism for this avirulence has been a subject of debate. *galE* mutants are phenotypically rough (i.e., produce incomplete lipopolysaccharide [LPS]) when grown in the absence of galactose but can be rendered smooth (i.e., produce complete LPS) and therefore invasive by growth in the presence of low concentrations of galactose (Nikaido 1961). *galE* strains are sensitive to high concentrations of galactose because accumulation of phosphorylated galactose leads to the lysis of cells (Fukasawa and Nikaido 1959, 1961). It is uncertain whether the sensitivity to galactose (resulting in cell lysis) or the inability to synthesize complete LPS (which renders the cells susceptible to nonspecific host defense mechanisms) leads to avirulence *in vivo*. Nevertheless, most *galE* mutants are difficult to grow in the presence of galactose without accumulating galactose-resistant mutants (Nikaido 1961), which are totally avirulent and also nonimmunogenic (Germanier and Furer 1971). In addition, the *galE* mutants are difficult to grow and lyophilize in a state to enable reconstitution in a highly immunogenic form (Hone et al. 1987). *galE* mutants of *S. choleraesuis* retain virulence, most likely because of lower requirements for galactose to produce complete LPS (Nnalue and Stocker 1986). Furthermore, Gal^+ revertants of the *S. typhi* Ty21a vaccine strain remained avirulent for mice, an observation which suggests that one or more additional mutations are contributing to the avirulence of the vaccine strain (Silva-Salinas et al. 1985). Similar observations have been made with an *ΔaroA S. typhimurium* vaccine strain which also seems to have accumulated additional mutation(s) conferring avirulence since restoration with the wild-type $aroA^+$ allele did not restore virulence (Lockman and Curtiss 1988). A more troublesome concern was the finding that certain genetically-engineered *galE* strains retained virulence (Hone et al. 1988).

Although not fully investigated, a mutation abolishing phosphomannose isomerase which also leads to a reversible rough/smooth phenotype dependent upon growth in the presence of mannose might also be useful for attenuating *Salmonella* but retaining immunogenicity.

Temperature-Sensitive Mutants

Morris Hooke et al. (1985) have isolated temperature-sensitive (Ts) "coaster" mutants of *Pseudomonas aeruginosa* and more recently, of other pathogens. These Ts "coaster" mutants grow more slowly at body temperature than wild-type strains and therefore, are presumably unable to multiply fast enough to cause disease prior to the time that protective immunity is induced (Morris Hooke et al. 1987; Sordelli et al. 1987). The fact that Ts defects are most likely due to missense mutations to result in amino acid substitutions to confer thermal sensitivity of a protein is a concern since such mutations should be capable of reversion. Even though the mutations were induced by nitrosoguanidine, they revert at frequencies of 10^{-6} to 10^{-7} (Morris Hooke et al. 1987). Since safety of vaccine strains is essential, inclusion of two separate mutations conferring the Ts "coaster" phenotype would be advisable and should diminish the probability for reversion. It is also of some concern that the products of the genes mutated to yield the Ts "coaster" phenotype are as yet unknown. Ohta et al. (1987) have tried a similar approach with *S. enteritidis*. Their mutants had UV-induced mutations, but the one that failed to revert unfortunately did not induce the highest level of immunity.

Other Mutant Phenotypes

After having worked with most of the mutant types mentioned above, we chose to explore the existence of mutations that would render *Salmonella* avirulent, yet immunogenic, and have the following properties: 1) be deletions, and therefore, unable to revert; 2) be two or more in number to preclude the likelihood that the phenotypic property could be lost by any gene transfer procedure; and 3) not be phenotypically reversed by anything in the diet or anything that could be supplied by the mammalian host. After considering many types of mutational lesions that might satisfy these criteria, it became apparent that many of the genes of interest were subject to catabolite repression; that is, their transcription was totally or partially dependent upon the presence of cyclic AMP (cAMP) which interacts with the cAMP receptor protein (CRP) as a positive activator of transcription (Pastan and Adhya 1976). More specifically, cAMP and the CRP protein are needed for the expression of numerous genes for carbohydrate and amino acid transport and/or utilization (reviewed by Rickenberg 1974; Pastan and Adhya 1976; Alper and Ames 1978; Botsford 1981), amino acid synthesis (Pastan and Perlman 1970; Prusiner et al. 1972), glycogen synthesis (Dietzler et al. 1977, 1979), synthesis of type I pili (Harwood and Meynell 1975; Saier et al. 1978), synthesis of flagella (Yokota and Gots 1970; Komeda et al. 1975), and synthesis of outer membrane proteins (Movva et al. 1981; Bremer et al. 1988). We therefore isolated, constructed, and tested *Δcya*, *Δcrp*, and *Δcya Δcrp S. typhimurium* strains with and without the pStSR100 virulence plasmid (Curtiss and Kelly 1987). Mice orally fed 10,000 times what would normally be an LD_{50} dose of the wild-type parent were unaffected and acquired immunity to challenge with 10,000 times a wild-type lethal dose by the oral route (Curtiss and Kelly 1987). Such immunity could be induced with a single immunization with as few as 1×10^7 *Δcya Δcrp S. typhimurium* cells (Curtiss et al. 1988b) and was demonstrable for up to three months (Curtiss et al. 1988a). Whereas oral inoculation of *ΔaroA S. typhimurium* strains results in significant titers in the GALT, liver, and spleen (Clements et al. 1986; Maskell et al. 1987), *Δcya Δcrp* strains possessing the pStSR100 virulence plasmid attach to, invade, and persist in the GALT, but are found in reduced titers in the spleen (Curtiss and Kelly 1987). Thus, the *Δcya Δcrp* mutants seem to be impaired in either reaching the spleen or in surviving once they get there. During *in vitro* growth the *Δcya Δcrp* strains have generation times 50% longer than their wild-type parents (Curtiss and Kelly 1987) and this may contribute to their avirulence.

Because both the *Δcya* and *Δcrp* mutations are deletions they cannot revert. However, suppressor mutations can occur rarely in *Δcrp* and *Δcya Δcrp* strains, usually affecting transcription of a single gene or operon (Scholte and Postma 1980; Melton et al. 1981; Garges and Adhya 1985). Such suppressor mutants so far tested in our *Δcya Δcrp* strains retain their avirulence and immunogenicity. Transduction of *Δcya* and *Δcrp* mutations to *cya*$^+$ or *crp*$^+$ occurs at reduced frequencies due to the extent of the deletions. Since the *Δcya* and *Δcrp* mutations are eleven minutes apart on the *Salmonella* chromosome (Sanderson and Hurley 1987), they could only be restored to the wild-type genotype in a single step by mating with an Hfr donor and Hfr donors, to the best of our knowledge, have not been observed in either *E. coli* or *Salmonella* strains isolated in nature. The use of the vaccine strain lacking the pStSR100 virulence plasmid further contributes to safety because this plasmid is not self-transmissible and presumably could only be reacquired by the vaccine strain mating with a *Salmonella* strain possessing a conjugative plasmid to mobilize the virulence plasmid (Jones et al. 1982; Gulig and Curtiss 1987).

RECOMBINANT BIVALENT AVIRULENT *SALMONELLA* VACCINE STRAINS

A diversity of recombinant avirulent strains have been constructed using either classical means of gene transfer or recombinant techniques to express a diversity of colonization and virulence antigens. Table 2 lists many of these constructs wherein genetic information for colonization and/or virulence antigens for species in the *Enterobacteriaceae* have been expressed in avirulent *Salmonella* strains. In addition, a number of colonization and/or virulence antigens of streptococcal pathogens have been expressed in avirulent *S. typhimurium* strains (Table 2) as have protein antigens of *Mycobacterium leprae*. Genes for *M. leprae* antigens were identified in recombinant gene libraries screened with patient sera (Mundayoor, Clark-Curtiss and Curtiss, unpublished results). Recombinant avirulent *S. typhimurium* strains and *S. typhi* Ty21a constructs have been used to immunize mice (Stevenson and Manning 1985; Clements et al. 1986; Curtiss et al. 1987, 1988a, 1988b; Maskell et al. 1986, 1987; Dougan et al. 1987a, 1987b) and humans (Tramont et al. 1984; Black et al. 1987), respectively. Secretory IgA and humoral IgA and IgG against the colonization or virulence antigen expressed by the recombinant *Salmonella* strain have usually been detected (see Table 2). On the other hand, induction of protective immunity against the pathogen supplying the colonization and/or virulence antigens has yet to be demonstrated. The principle reasons are undoubtedly due to the following: 1) most of the recombinant strains were unable to stably maintain the cloned gene on a plasmid vector and thus, the ability to express the cloned gene was lost as a function of time after colonization of the GALT; 2) many of the strains failed to produce the foreign colonization or virulence antigen to the extent observed when the cloned gene was in typical *E. coli* K-12 cloning hosts; 3) in the case of *S. typhimurium* strains with Δ*asd* mutations, the strains did not persist long enough in the GALT to stimulate an adequate immune response; and 4) the foreign antigen may not be capable of inducing a protective immune response.

Our results with the Δ*cya* Δ*crp* *S. typhimurium* strains have indicated that at least the first three of these potential pitfalls do not pertain. First, these strains express cloned gene products at levels significantly higher than in *E. coli* strains, even *E. coli* strains with Δ*cya* mutations (Nakayama and Curtiss, unpublished results). Second, several recombinant plasmids expressing foreign colonization antigens are stably maintained, such that 80-90% of the recombinant *Salmonella* recovered from murine GALT eight days after oral inoculation, still possess the recombinant plasmid and express the cloned foreign colonization antigen (Curtiss et al. 1988a). The molecular bases for these beneficial attributes of the Δ*cya* Δ*crp* strains are as yet not understood.

IMPROVED VECTORS FOR USE WITH AVIRULENT *SALMONELLA* VACCINE STRAINS

The Food and Drug Administration is loathe to approve live vaccine strains that express antibiotic resistance, thus, use of cloning vectors with drug resistance markers is to be avoided. In addition, selective maintenance of the recombinant clone *in vivo* would be difficult to achieve using plasmid cloning vectors conferring drug resistance. We therefore have developed a vector-host system that constitutes a "balanced lethal."

Table 2. Recombinant avirulent *Salmonella* to induce immune responses

Salmonella mutant	Expressed foreign antigen	Type of immune response	References
S. typhi Ty21a *galE*	*Shigella sonnei* O-antigen	secretory & humoral	Formal et al. (1981) Tramont et. al. (1984) Black et al. (1987)
S. typhimurium *galE*	*E. coli* K88 fimbriae	secretory & humoral	Stevenson and Manning (1985)
S. enteritidis Δ*aroA*	B subunit of *E. coli* enterotoxin	secretory & humoral	Clements et al. (1986)
S. typhi Ty21a *galE*	*Streptococcus sobrinus* surface protein antigen (SpaA) and *Streptococcus mutans* glucosyltransferase (GtfA)	N.I.[a]	Curtiss et al. (1986)
S. typhimurium Δ*asd*, Δ*aroA*	*S. sobrinus* SpaA *S. mutans* GtfA	secretory & humoral	Curtiss (1986) Curtiss et al. (1986, 1987)
S. typhimurium Δ*aroA*	*E. coli* K88 fimbriae	humoral	Dougan et al. (1986)
S. typhimurium Δ*aroA*	*E. coli* LT-B subunit	secretory & humoral	Maskell et al. (1987)
S. typhimurium Δ*aroA*	*E. coli* β-galactosidase	humoral & cellular	Brown et al. (1987)
S. typhi Ty21a *galE*	*Shigella flexneri* 2a Type & group antigens	humoral	Baron et al. (1987)
S. typhimurium Δ*cya* Δ*crp*	*Streptococcus sobrinus* SpaA	U.I.[b]	Curtiss et al. (1988a)
S. typhimurium Δ*cya* Δ*crp* Δ*asd*	*S. sobrinus* SpaA	U.I.[b]	Nakayama et al. (1988)
S. typhimurium Δ*cya* Δ*crp*	*Streptococcus equi* M. protein	U.I.[b]	Curtiss et al. (1988b) Galan et al. (1988)
S. typhimurium Δ*cya* Δ*crp* Δ*asd*	*Mycobacterium leprae* surface protein antigens	U.I.[b]	Mundayoor, Clark-Curtiss, and Curtiss, unpublished

[a] Not investigated
[b] Under investigation

This was accomplished with a mutation in the chromosome blocking synthesis of an essential metabolite that is not readily available in nature, and certainly not in the animal host, and with a plasmid vector containing a non-homologous gene sequence complementing the chromosomal gene defect (Nakayama et al. 1988). Diaminopimelic acid (DAP) is a unique constituent of the rigid layer of the bacterial cell wall in gram-negative bacteria (Schleifer and Kandler 1972). The inability to synthesize DAP when growth is possible results in "DAPless death" associated with cell lysis and release of cell contents (Meadow and Work 1956; Bauman and Davis 1957; Meadow et al. 1957; Rhuland 1957). DAP is only synthesized by bacteria and a few plants. We have thus made use of *E. coli* and *Salmonella* strains with a deletion for the gene for aspartate β-semialdehyde dehydrogenase (*asd*), which thus have requirements for threonine, methionine and DAP, and plasmid vectors that contain the *asd*+ gene cloned from *Streptococcus mutans* (Jagusztyn-Krynicka et al. 1982). The *Streptococcus mutans asd* gene is expressed at very high levels in *E. coli* (Jagusztyn-Krynicka et al. 1982; Cardineau and Curtiss 1987) and *S. typhimurium* (Nakayama and Curtiss, unpublished results) due to the presence of a most unusual promoter that has a very high affinity for RNA polymerase (Cardineau and Curtiss 1987). A map of the Asd+ cloning vector pYA248 is depicted in Figure 1. This vector has the *trc* promoter, a multiple cloning site, the *asd* promoter and structural gene, and transcription terminators derived from the *rrnB* gene (Amann and Brosius 1985). The introduction of pYA248 in *S. typhimurium Δasd* mutant restores wild-type virulence in mouse infectivity studies. A *Δcya Δcrp Δasd S. typhimurium* strain (χ4072) has been constructed to use with the pYA248 vector into which we have cloned genes for colonization and virulence antigens from a number of pathogens. The cloning of the *lacZ* gene specifying β-galactosidase into pYA248 and introduction of this construct into χ4072 results in a strain that constitutively synthesizes approximately five-times more β-galactosidase than is expressable from *E. coli* strains. In addition, χ4072 with the *Streptococcus sobrinus spaA* gene cloned into pYA248 similarly produces more SpaA protein that is highly antigenic than either *S. typhimurium ΔaroA* strains or *E. coli* cloning hosts (Nakayama et al. 1988). More important, maintenance of these strains in the absence of DAP results in 100% of the population continuing to express the cloned gene, whether growth is *in vitro* or *in vivo*. When DAP is present in growth medium, approximately 1% of the bacteria per generation lose the plasmid and undergo DAPless death upon subsequent growth in the absence of DAP. If this were to occur *in vivo*, the lysis of bacteria with liberation of their contents should serve to augment stimulation of the immune response.

The use of the Asd+ vectors in *Δcya Δcrp Δasd Salmonella* strains constitutes at present a very promising host-vector system for stimulating a variety of immune responses. These strains are genotypically stable, are completely avirulent, are highly immunogenic, and the gene defects in the strains are not remediable by diet or by the host. In addition, the strains are easy to grow and store, and recombinant plasmids are maintained with high stability and confer a high level of expression of cloned gene products.

SUMMARY

The means to attenuate *Salmonella* and to endow such avirulent strains with the ability to express colonization and virulence antigens from other pathogens has achieved considerable progress during the past several years. One can therefore begin to design and construct strains with specificity to a given animal host and to express in a defined way specific colonization and virulence antigens in a manner to stimulate long-lasting immunity to the *Salmonella* and to the pathogen supplying the genetic information for the colonization and virulence antigens. Since most pathogens colonize on or invade through mucosal surfaces, the use of recombinant bivalent *Salmonella* vaccine strains to stimulate a mucosal immune response would induce the development of a first line of defense against a diversity of pathogens. Mucosal immunity should therefore reduce contagious spread of many pathogens since the dose to overcome the mucosal immune barrier would be increased to result in a diminished likelihood of infection. The fact that the recombinant *Salmonella* vaccine strains also induce humoral and cellular immune responses justifies their use for induction of long-lasting immunity.

Although considerable progress has been made in targeting antigens to the GALT by use of avirulent *Salmonella*, a similar strategy for delivery of antigens to the BALT has yet to be discovered and developed.

In addition to constituting a system for induction of immunity against a diversity of pathogens, the recombinant avirulent *Salmonella* system should provide a means to explore parameters of the mucosal immune response. This would include investigation of the location and duration of memory, the age dependence of induction of mucosal immunity, and the means for the possible induction of oral tolerance with regard to either the mucosal or humoral response to an antigen expressed by the recombinant *Salmonella*. It is also possible to contemplate using the avirulent *Salmonella* to target expression of various modulators of the immune system such as interleukin-2 and interferon-γ to the GALT and thus further enhance the immune response.

Lastly, one can introduce into avirulent *Salmonella* strains genes for putative colonization antigens in order to investigate whether induction of an immune response against the putative colonization antigen does or does not interfere with infection. This system, therefore, permits another means to analyze the relative importance of various bacterial surface attributes in conferring pathogenicity to the microbe.

Acknowledgments. We thank Kathleen Eaton and Ruth Lewis for help in preparation of this manuscript, and Jorge E. Galan and Hank A. Lockman for consultation.

REFERENCES

Alper MD, Ames BN (1978) Transport of antibiotics and metabolite analogs by systems under cyclic AMP control: Positive selection of *Salmonella typhimurium cya* and *crp* mutants. J Bacteriol 133:149-157

Amann E, Brosius J (1985) 'ATG vectors' for regulated high-level expression of cloned genes in *Escherichia coli*. Gene 40:183-190

Asherson GL, Zembala M, Perera MACC, Mayhew B, Thomas WR (1977) Production of immunity and unresponsiveness in the mouse by feeding contact sensitizing agents and the role of suppressor cells in the Peyer's patches, mesenteric lymph nodes, and other lymphoid tissues. Cell Immunol 33:145-155

Bacon GA, Burrows TW, Yates M (1950) The effects of biochemical mutation on the virulence of *Bacterium typhosum*: The virulence of mutants. Brit J Exp Path 32:714-724

Bacon GA, Burrows TW, Yates M (1951) The effects of biochemical mutation on the virulence of *Bacterium typhosum*: The loss of virulence of certain mutants. Brit J Exp Pathol 32:85-96

Baron LS, Kopecko DJ, Formal SB, Seid R, Guerry P, Powell C (1987) Introduction of *Shigella flexneri* 2a type and group antigen genes into oral typhoid vaccine strain *Salmonella typhi* Ty21a. Infect Immun 55:2797-2801

Barrow PA, Simpson JM, Lovell MA, Binns MM (1987) Contribution of *Salmonella gallinarum* large plasmid toward virulence in fowl typhoid. Infect Immun 55:388-392

Bauman N, Davis BD (1957) Selection of auxotrophic bacterial mutants through diaminopimelic acid or thymine deprival. Science 126:170

Bienenstock J, McDermott M, Befus D, O'Neil LM (1978) A common mucosal immunologic system involving the bronchus, breast, and bowel. Adv Exp Med Biol 107:53-59

Black RE, Levine MM, Clements ML, Losonsky G, Herrington D, Berman S, Formal SB (1987) Prevention of shigellosis by a *Salmonella typhi- Shigella sonnei* bivalent vaccine. J Infect Dis 155:1260-1265

Bochner BR, Huang HC, Schieven GL, Ames BN (1980) Positive selection for loss of tetracycline resistance. J Bacteriol 143:926-933

Botsford JL (1981) Cyclic nucleotides in prokaryotes. Microbiol Rev 45:620-642

Bremer E, Gerlach P, Middendorf A (1988) Double negative and positive control of *tsx* expression in *Escherichia coli*. J Bacteriol 170:108-116

Brown A, Hormaeche CE, Demarco de Hormaeche R, Winther M, Dougan G, Maskell DJ, Stocker BAD (1987) An attenuated *aroA Salmonella typhimurium* vaccine elicits humoral and cellular immunity to cloned β-galactosidase in mice. J Infect Dis 155:86-92

Cardineau GA, Curtiss III R (1987) Nucleotide sequence of the *asd* gene of *Streptococcus mutans*: Identification of the promoter region and evidence for attenuator-like sequences preceding the structural gene. J Biol Chem 262:3344-3353

Carter PB, Collins FM (1974) The route of enteric infection in normal mice. J Exp Med 139:1189-1203

Cebra JJ, Gearhart PJ, Kamat R, Robertson SM, Tseng J (1976) Origin and differentiation of lymphocytes involved in the secretory IgA response. Cold Spring Harbor Symp Quant Biol 41:201-221

Chikami GK, Fierer J, Guiney DG (1985) Plasmid-mediated virulence in *Salmonella dublin* demonstrated by use of a Tn*5-oriT* construct. Infect Immun 50:420-424

Clements JD (1987) Use of attenuated mutants of *Salmonella* as carriers for delivery of heterologous antigens to the secretory immune system. Pathol Immunopathol Res 6:137-146

Clements JD, El-Morshidy S (1984) Construction of a potential live oral bivalent vaccine for typhoid fever and cholera-*Escherichia coli*-related diarrheas. Infect Immun 46:564-569

Clements JD, Lyon FL, Lowe KL, Farrand AL, El-Morshidy S (1986) Oral immunization of mice with attenuated *Salmonella enteritidis* containing a recombinant plasmid which encodes for production of the B subunit of heat-labile *Escherichia coli* enterotoxin. Infect Immun 53:685-692

Cohen SN, Chang ACY, Boyer HW, Helling RB (1973) Construction of biologically functional bacterial plasmids *in vitro*. Proc Natl Acad Sci USA 70:3240-3244

Collins FM (1972) Salmonellosis in orally infected specific pathogen-free C57BL mice. Infect Immun 5:191-198

Curtiss III R (1978) Biological containment and cloning vector transmissibility. J Infect Dis 137:668-675

Curtiss III R (1985) Genetic analysis of *Streptococcus mutans* virulence. Curr Top Microbiol Immun 118:253-277

Curtiss III R (1986) Genetic analysis of *Streptococcus mutans* virulence and prospects for an anticaries vaccine. J Dent Res 65:1034-1045

Curtiss III R, Kelly SM (1987) *Salmonella typhimurium* deletion mutants lacking adenylate cyclase and cyclic AMP receptor protein are avirulent and immunogenic. Infect Immun 55:3035-3043

Curtiss III R, Pereira DA, Hsu JC, Hull SC, Clark JE, Maturin Sr LJ, Goldschmidt R, Moody R, Inoue M, Alexander L (1976) Biological containment: The subordination of *Escherichia coli* K-12. In: Beers Jr RF, Bassett EG (eds) Recombinant Molecules: Impact on Science and Society. Raven Press New York NY, p 45

Curtiss III R, Inoue M, Pereira D, Hsu JC, Alexander L, Rock L (1977) Construction and use of safer bacterial host strains for recombinant DNA research. In: Scott WA, Werner RA (eds) Molecular Cloning of Recombinant DNA. Academic Press New York, p 99

Curtiss III R, Holt RG, Barletta R, Robeson JP, Saito S (1983) *Escherichia coli* strains producing *Streptococcus mutans* proteins responsible for colonization and virulence. In: McGhee JR, Mestecky J (eds) The Secretory Immune System, Vol. 409. Ann NY Acad Sci p 688

Curtiss III R, Goldschmidt R, Pastian R, Lyons M, Michalek SM, Mestecky J (1986) Cloning virulence determinants from *Streptococcus mutans* and the use of recombinant clones to construct bivalent oral vaccine strains to confer protective immunity against *S. mutans*-induced dental caries. In: Hamada S, Michalek SM, Kiyono H, Menaker L, McGhee JR (eds) Molecular Microbiology and Immunology of *Streptococcus mutans*. Elsevier North Holland p 173

Curtiss III R, Goldschmidt R, Kelly SM, Lyons M, Michalek S, Pastian R, Stein S (1987) Recombinant avirulent *Salmonella* for oral immunization to induce mucosal immunity to bacterial pathogens. In: Kohler H, LoVerde PT (eds) Vaccines: New Concepts and Developments. Proceedings of the Tenth International Convocation on Immunology. Longman Scientific and Technical, Harlow, Essex p 261

Curtiss III R, Goldschmidt RM, Fletchall NB, Kelly SM (1988a) Avirulent *Salmonella typhimurium Δcya Δcrp* oral vaccine strains expressing a streptococcal colonization and virulence antigen. Vaccine 6:155-160

Curtiss III R, Kelly SM, Gulig PA, Gentry-Weeks CR, Galan JE (1988b) Avirulent *Salmonella* expressing virulence antigens from other pathogens for use as orally-administered vaccines. In: Roth J (ed) Virulence Mechanisms. American Society for Microbiology, Washington DC, p 311

Dietzler DN, Leckie MP, Sternheim WL, Taxman TL, Unger JM, Porter SE (1977) Evidence for the regulation of bacterial glycogen synthesis by cyclic AMP. Biochem Biophys Res Commun 77:1468-1477

Dietzler DN, Leckie MP, Magnini JL, Sughrue MJ, Bergstein PE, Sternheim WL (1979) Contribution of adenosine cyclic 3':5'-monophosphate to the regulation of bacterial glycogen synthesis *in vivo*. J Biol Chem 254:8308-8317

Dougan G, Sellwood R, Maskell D, Sweeney K, Liew FY, Beesley J, Hormaeche C (1986) *In vivo* properties of a cloned K88 adherence antigen determinant. Infect Immun 52:344-347

Dougan G, Hormaeche CE, Maskell DJ (1987a) Live oral *Salmonella* vaccines: Potential use of attenuated strains as carriers of heterologous antigens to the immune system. Parasite Immunol 9:151-160

Dougan G, Maskell D, Pickard D, Hormaeche C (1987b) Isolation of stable *aroA* mutants of *Salmonella typhi* Ty2: Properties and preliminary characterization in mice. Mol Gen Genet 207:402-405

Elson CO, Ealding W (1984) Generalized systemic and mucosal immunity in mice after mucosal stimulation with cholera toxin. J Immunol 132:2736-2741

Formal SB, Baron LS, Kopecko DJ, Powell C, Life CA (1981) Construction of a potential bivalent vaccine strain: Introduction of *S. sonnei* form I antigen genes into the *galE Salmonella typhi* Ty21a typhoid vaccine strain. Infect Immun 34:746-750

Fukasawa T, Nikaido H (1959) Galactose-sensitive mutants of *Salmonella*. Nature (London) 184:1168-1169

Fukasawa T, Nikaido H (1961) Galactose-sensitive mutants of *Salmonella*. II. Bacteriolysis induced by galactose. Biochim Biophys Acta 48:470-483

Gaines S, Sprinz H, Tully JG, Tigertt WD (1968) Studies on infection and immunity in experimental typhoid fever. VII. The distribution of *Salmonella typhi* in chimpanzee tissue following oral challenge and the relationship between the numbers of bacilli and morphologic lesions. J Infect Dis 118:293-306

Galan JE, Timoney JF, Curtiss III R (In press) Expression and localization of the *Streptococcus equi* M protein in *Escherichia coli* and *Salmonella typhimurium*. In: Powell D (ed) Proceeding of the Fifth International Congress on Equine Infectious Diseases. Kentucky University Press, Lexington, KY

Garges S, Adhya S (1985) Sites of allosteric shift in the structure of the cyclic AMP receptor protein. Cell 41:745-751

Germanier R, Furer E (1971) Immunity in experimental salmonellosis. II. Basis for the avirulence and protective capacity of *galE* mutants of *Salmonella typhimurium*. Infect Immun 4:663-673

Germanier R, Furer E (1975) Isolation and characterization of GalE mutant Ty21a of *Salmonella typhi*: A candidate strain for a live, oral typhoid vaccine. J Infect Dis 131:553-558

Goebel W (ed) (1985) Genetic approaches to microbial pathogenicity. In: Curr Top Microbiol Immun, Vol. 118, Springer-Verlag New York

Gulig PA, Curtiss III R (1987) Plasmid-associated virulence of *Salmonella typhimurium*. Infect Immun 55:2891-2901

Gulig, PA, Curtiss III R (In press) Transposon-insertion mutagenesis of cloned virulence genes of the 100-kilobase plasmid of *Salmonella typhimurium*. Infect Immun

Habasha FG, Smith BP, Schwartz L, Ardans A, Reina-Guerra M (1985) Correlation of macrophage migration-inhibition factor and protection from challenge by exposure in calves vaccinated with *Salmonella typhimurium*. Am J Vet Res 46:1415-1421

Hackett J, Kotlarski IK, Mathan V, Francki K, Rowley D (1986) The colonization of Peyer's patches by a strain of *Salmonella typhimurium* cured of the cryptic plasmid. J Infect Dis 153:1119-1125

Hackett J, Wyk P, Reeves P, Mathan V (1987) Mediation of serum resistance in *Salmonella typhimurium* by an 11-kilodalton polypeptide encoded by the cryptic plasmid. J Infect Dis 155:540-549

Harwood CR, Meynell E (1975) Cyclic AMP and the production of sex pili by *E coli* K-12 carrying depressed sex factors. Nature 254:628-630

Heffernan EJ, Fierer J, Chikami G, Guiney D (1987) Natural history of oral *Salmonella dublin* infection in BALB/c mice: Effect of an 80-kilobase-pair plasmid on virulence. J Infect Dis 155:1254-1259

Hoiseth S, Stocker BAD (1981) Aromatic-dependent *Salmonella typhimurium* are nonvirulent and effective as live vaccines. Nature 291:238-239

Hone DM, Attridge SR, Forrest B, Morona R, Daniels D, LaBrooy JT, Bartholomeusz RCA, Shearman DJC, Hackett J (1988) A *galE via* (Vi Antigen-Negative) Mutant of *Salmonella typhi* Ty2 Retains Virulence in Humans. Infect Immun 56:1326-1333

Hone D, Morona R, Attridge S, Hackett J (1987) Construction of defined *galE* mutants of *Salmonella* for use as vaccines. J Infect Dis 156:167-174

Jagusztyn-Krynicka EK, Smorawinska M, Curtiss III R (1982) Expression of *Streptococcus mutans* aspartate semialdehyde dehydrogenase gene cloned into plasmid pBR322. J Gen Microbiol 128:1135-1145

Jones GW, Rabert DK, Svinarich DM, Whitfield HJ (1982) Association of adhesive, invasive, and virulent phenotypes of *Salmonella typhimurium* with autonomous 60-megadalton plasmids. Infect Immun 338:476-486

Killar LM, Eisenstein TM (1985) Immunity to *Salmonella typhimurium* infection in C3H/HeJ and C3H/HeNCr1BR mice: Studies with an aromatic-dependent live *Salmonella typhimurium* strain as a vaccine. Infect Immun 47:605-612

Kleckner N, Roth J, Botstein D (1977) Genetic engineering *in vivo* using translocatable drug-resistance elements. J Mol Biol 116:125-159

Komeda Y, Suzuki H, Ishidsu J, Iino T (1975) The role of cAMP in flagellation of *Salmonella typhimurium*. Mol Gen Genet 142:289-298

LeFever ME, Joel DD (1984) Peyer's patch epithelium: An imperfect barrier. In: Schiller CM (ed) Intestinal Toxicology. Raven Press New York, p 45

Levine MM, Herrington D, Murphy JR, Morris JG, Losonsky G, Tall B, Lindberg AA, Svenson S, Baqar S, Edwards MF, Stocker B (1987) Safety, infectivity, immunogenicity, and *in vivo* stability of two attenuated auxotrophic mutant strains of *Salmonella typhi*, 541Ty and 543Ty, as live oral vaccines in humans. J Clin Invest 79:888-902

Levy E, Gaehtgens W (1908) Uber die verbreitung der typhusbazillen in den lymphdrusen bei typhusleichen. Arb Kaiser Gesundh 28:168-171

Lindberg AA, Robertsson JA (1983) *Salmonella typhimurium* infection in calves: Cell-mediated and humoral immune reactions before and after challenge with live virulent bacteria in calves given live or inactivated vaccines. Infect Immun 41:751-757

Lockman HA, Curtiss III R (In press) Complementation and selection *in vivo* for cloned genes restoring mouse virulence to attenuated *Salmonella typhimurium*. Abstracts Annual Meeting Amer Soc Microbiol

Macrina FL (1984) Molecular cloning of bacterial antigens and virulence determinants. Annu Rev Microbiol 38:193-219

Maloy SR, Nunn WD (1981) Selection for loss of tetracycline resistance by *Escherichia coli*. J Bacteriol 145:1110-1112

Manning EJ, Baird GD, Jones PW (1986) The role of plasmid genes in the pathogenicity of *Salmonella dublin*. J Med Microbiol 21:239-243

Maskell D, Liew FY, Sweeney K, Dougan G, Hormaeche C (1986) Attenuated *Salmonella typhimurium* as live oral vaccines and carriers for delivering antigens to the secretory immune system. In: Brown F, Chanock RM, Lerner RA (eds) Vaccines 86: New approaches to immunization developing vaccines against parasitic bacterial and viral diseases. Cold Spring Harbor Laboratory, Cold Spring Harbor, p 213

Maskell D, Sweeney KJ, O'Callaghan D, Hormaeche CE, Liew FY, Dougan G (1987) *Salmonella typhimurium aroA* mutants as carriers of the *Escherichia coli* heat-labile enterotoxin B subunit to the murine secretory and systemic immune systems. Microb Pathogen 2:211-221

McCaughan G, Basten A (1983) Immune system of the gastrointestinal tract. Int Rev Physiol 28:131-157

McFarland WC, Stocker BAD (1987) Effect of different purine auxotrophic mutations on mouse-virulence of a Vi-positive strain of *Salmonella dublin* and of two strains of *Salmonella typhimurium*. Microb Pathogen 3:129-141

Meadow P, Hoare DS, Work E (1957) Interrelationships between lysine and αε-diaminopimelic acid and their derivatives and analogues in mutants of *Escherichia coli*. Biochemical Journal 66:270-282

Meadow P, Work E (1956) Interrelationships between diaminopimelic acid, lysine and their analogues in mutants of *Escherichia coli*. Biochemical Journal 64:11p

Melton T, Snow LL, Freitag CS, Dobrogosz WJ (1981) Isolation and characterization of cAMP suppressor mutants of *Escherichia coli* K-12. Mol Gen Genet 182:480-489

Michiels T, Popoff MY, Durviaux S, Coynault C, Cornelis G (1987) A new method for the physical and genetic mapping of large plasmids: Application to the localization of the virulence determinants on the 90 kb plasmid of *Salmonella typhimurium*. Microb Pathogen 3:109-116

Morris Hooke A, Sordelli DO, Cerquetti MC, Bellanti JA (1987) Differential growth characteristics and immunogenicity of tight and coasting temperature-sensitive mutants of *Pseudomonas aeruginosa*. Infect Immun 55:99-103

Morris Hooke A, Bellanti JA, Oeschger MP (1985) Live bacterial vaccines: New approaches for safety and efficacy. Lancet i:1472-1474

Movva RN, Green P, Nakamura K, Inouye M (1981) Interaction of cAMP receptor protein with the *ompA* gene, a gene for a major outer membrane protein of *Escherichia coli*. FEBS Letters 128:186-190

Mukkur TKS, McDowell GH, Stocker BAD, Lascelles AK (1987) Protection against experimental salmonellosis in mice and sheep by immunization with aromatic-dependent *Salmonella typhimurium*. J Med Microbiol 24:11-19

Muller M (1912) Der nachweis von fleischvergiftungsbakterien in fleisch und organen von schlachttieren auf grund systematischer untersuchungen uber den verlauf und den mechanismus der infektion des tierkorpers mit bakterien der enteritis- und Paratyphus-gruppe, sowie des Typhus; Zugleich ein beitrag zum infektions- und virulenzproblem der bakterien auf experimenteller basis. Centroblatt Bakteriolie Parasitenkunde 62:335-373

Nakamura SY, Sato S, Ohya T, Suzuki S, Ikeda S (1985a) Possible relationship of a 36-megadalton *Salmonella enteritidis* plasmid to virulence in mice. Infect Immun 47:831-833

Nakamura M, Sato S, Ohya T, Suzuki S, Ikeda S, Koeda T (1985b) Plasmid-cured *Salmonella enteritidis* AL1192 as a candidate for a live vaccine. Infect Immun 50:586-587

Nakayama K, Kelly SM, Curtiss III R (1988) Construction of an Asd^+ expression-cloning vector: Stable maintenance and high level expression of cloned genes in a *Salmonella* vaccine strain. Bio/Technology 6:693-697

Nikaido H (1961) Galactose-sensitive mutants of *Salmonella*. I. Metabolism of galactose. Biochim Biophys Acta 48:460-469

Nnalue NA, Stocker BAD (1986) Some *galE* mutants of *Salmonella choleraesuis* retain virulence. Infect Immun 54:635-640

Ohta M, Kido N, Fuji Y, Arakawa Y, Komatsu T, Kato N (1987) Temperature-sensitive growth mutants as live vaccines against experimental murine salmonellosis. Microbiol Immunol 31:1259-1265

Pardon P, Popoff MY, Coynault C, Marly J, Miras I (1986) Virulence-associated plasmids of *Salmonella* serotype *typhimurium* in experimental murine infection. Ann Microbiol (Inst Pasteur) 137B:47-60

Pastan I, Adhya S (1976) Cyclic adenosine 5'-monophosphate in *Escherichia coli*. Bacteriol Rev 40:527-551

Pastan I, Perlman R (1970) Cyclic adenosine monophosphate in bacteria. Science 169:339-344

Prusiner S, Miller RE, Valentine RC (1972) Adenosine 3':5'-Cyclic monophosphate control of the enzymes of glutamine metabolism in *Escherichia coli*. Proc Nat Acad Sci USA 69:2922-2926

Rickenberg HV (1974) Cyclic AMP in Prokaryotes. Annu Rev Microbiol 28:353-369

Robertsson JA, Lindberg AA, Hoiseth S, Stocker BAD (1983) *Salmonella typhimurium* infection in calves: Protection and survival of virulent challenge bacteria after immunization with live and inactivated vaccines. Infect Immun 41:742-750

Rhuland LE (1957) Role of αε-diaminopimelic acid in the cellular integrity of *Escherichia coli*. J Bacteriol 73:778-783

Saier MH, Schmidt MR, Leibowitz M (1978) Cyclic AMP-dependent synthesis of fimbriae in *Salmonella typhimurium*: Effects of *cya* and *pts* mutations. J Bacteriol 134:356-358

Sanderson KE, Hurley JA (1987) Linkage map of *Salmonella typhimurium*, In: Neidhardt FC (ed) *Escherichia coli and Salmonella typhimurium: Cellular and Molecular Biology* Edition VI, Vol. 2, American Society for Microbiology, Washington, D.C., p 877

Sanderson, KE, Ross H, Ziegler L, Makela PH (1972) F^+, Hfr, and F' strains of *Salmonella typhimurium* and *Salmonella abony*. Bacteriol Rev 36:608-637

Schleifer KH, Kandler, O (1972) Peptidoglycan types of bacterial cell walls and their taxonomic implications. Bacteriol Rev 36:407-477

Schmeiger H (1972) Phage P22 mutants with increased or decreased transduction abilities. Mol Gen Genet 119:75-88

Scholte BJ, Postma PW (1980) Mutation in the *crp* gene of *Salmonella typhimurium* which interferes with inducer exclusion. J Bacteriol 141:751-757

Silva-Salinas BA, Rodriguez-Aguayo L, Maldonado-Ballesteros A, Valenzuela-Montero ME, Seoane-Montecinos ME (1985) Propiedades de dos derivados Gal^+ de la cepa vacune *Salmonella typhi* gal E Ty21a. Bol Med Hosp Infant Mex 42:234-239

Smith BP, Reina-Guerra M, Hoiseth SK, Stocker BAD, Babasha F, Johnson E, Merritt F (1984) Aromatic-dependent *Salmonella typhimurium* as modified live vaccines for calves. Am J Vet Res 45:59-66

Sordelli DO, Cerquetti MC, Bellanti JA, Morris Hooke A (1987) Specific pulmonary defences against *Pseudomonas aeruginosa* after local immunization with temperature-sensitive mutants. J Gen Microbiol 133:2835-2841

Stevenson G, Manning PA (1985) Galactose epimeraseless (*galE*) mutant G30 of *Salmonella typhimurium* is a good potential live oral vaccine carrier for fimbrial antigens. FEMS Microbiol Letters 28:317-321

Terakado N, Sekizaki T, Hashimoto K, Naitoh S (1983) Correlation between the presence of a fifty-megadalton plasmid in *Salmonella dublin* and virulence for mice. Infect Imun 41:443-444

Tramont EC, Chung R, Berman S, Keren D, Kapfer C, Formal SB (1984) Safety and antigenicity of typhoid-*Shigella sonnei* vaccine (Strain 5076-1C). J Infect Dis 149:133-136

Weisz-Carrington P, Roux M, McWilliams M, Phillips-Quagliata JM, Lamm ME (1979) Organ and isotype distribution of plasma cells producing specific antibody after oral immunization: Evidence for a generalized secretory immune system. J Immunol 123:1705-1708

Yokota R, Gots JS (1970) Requirement of adenosine 3',5'-cyclic phosphate for flagella formation in *Escherichia coli* and *Salmonella typhimurium*. J Bacteriol 103:513-516

Liposomes as Oral Adjuvants

S.M. Michalek,[1] N.K. Childers,[2] J. Katz,[1] F.R. Denys,[2] A.K. Berry,[3] J.H. Eldridge,[1,3] J.R. McGhee[1], and R. Curtiss III[4]

[1]Department of Microbiology, [2]School of Dentistry and [3]Multipurpose Arthritis Center, The University of Alabama at Birmingham, Birmingham, AL 35294; and [4]Department of Biology, Washington University, St. Louis, MO 63130, USA

OVERVIEW OF THE COMMON MUCOSAL IMMUNE SYSTEM AND ORAL VACCINES

Studies in both humans and experimental animals have provided evidence for the existence of a common mucosal immune system, whereby oral administration of antigen results in the appearance of IgA antibodies in various external secretions (reviewed by Mestecky and McGhee 1987). IgA antibodies provide the primary defense mechanism against infections which involve the mucosal surfaces. Therefore, by better understanding the mechanisms involved in the common mucosal immune system and properties of oral vaccines important in inducing mucosal immune responses, we have the potential to protect the host against a variety of infectious diseases. In this brief review, we discuss factors important in developing both effective and safe oral vaccines. Emphasis will be placed on the recent approaches we have taken in the development of an oral vaccine, including the usefulness of liposomes as antigen carriers and as adjuvants, for the induction of a mucosal immune response protective against the infectious oral disease dental caries, in which the principal etiologic agent is *Streptococcus mutans* (reviewed by McGhee and Michalek 1981).

The Common Mucosal Immune System

The inductive events leading to IgA responses in external secretions occur in the gut-associated lymphoid tissue (GALT), e.g., Peyer's patches, which are distinct lymphoid follicles located along the small intestine. A specialized epithelium covering the luminal surface of Peyer's patches consists of cuboidal epithelial cells and unique antigen-sampling cells called M cells (microfolding and membranous cells; Owen and Jones 1974) or FAE cells (follicular-associated epithelial cells; Bockman and Cooper 1973). The M cells, which are actively pinocytotic and phagocytic, take up gut luminal antigens and deliver them to the underlying lymphoid cells located in the dome region of the Peyer's patch for antigen sensitization of T cells and precursor IgA B cells. These cells leave the Peyer's patch and migrate to mucosal tissues and secretory glands where the B cells differentiate into plasma cells secreting IgA antibodies with specificity for the antigen first encountered in GALT. This process, which is separate from the systemic immune mechanisms, has been termed the common mucosal immune system (Mestecky and McGhee 1987).

Current Topics in Microbiology and Immunology, Vol. 146

Characteristics of Oral Vaccines

A number of antigen forms have been used in the development of oral vaccines. Classical oral vaccines have consisted of killed, attenuated or living organisms, e.g., bacteria, parasites or viruses (reviewed by Mestecky 1987; Mestecky and McGhee 1987; and elsewhere in this book). These forms of vaccines possess a variety of antigenic determinants and adjuvant molecules which may be useful in demonstrating that a response can be induced. Whole cell vaccines may not, however, be the most practical since they may possess components which could induce detrimental effects in the host or they could induce a response that may not be effective in protection. Therefore, it would be useful to identify and purify the potential protective antigen. It has been shown that a vaccine consisting of the whole *S. mutans* cells is effective in inducing secretory immune responses when given orally to either humans (McGhee and Michalek 1981; Czerkinsky et al. 1987) or experimental animals (McGhee and Michalek 1981; Michalek et al. 1983). The induction of specific salivary IgA antibodies correlated with a reduction of *S. mutans* in plaque in both humans and animals and with protection against caries development in animals. The cell wall of *S. mutans* consists of a variety of antigens and adjuvant molecules of potential importance in caries immunity. Interspersed throughout the peptidoglycan layer and exposed on the surface of the cell are serotype-specific carbohydrate (CHO), polyglycerol phosphate, lipoteichoic acid and various surface proteins. These proteins are loosely associated with the cell surface and are released into the culture supernatant in relatively large quantities, which facilitates their isolation and purification. Antigens of *S. mutans*, which have been shown to induce protective immune responses include the CHO, glucosyltransferase, antigen I/II, as well as other protein antigens (reviewed in Hamada et al. 1986; Russell and Mestecky 1986).

Once the protective antigen(s) has been identified, it must be shown to be immunogenic in its purified form and via the selected route of delivery. In this regard, it has been demonstrated that soluble antigens are not as immunogenic as particulate forms when used as oral vaccines. Therefore, it may be necessary to use a carrier/vehicle or an adjuvant with a soluble antigen for inducing an adequate response (see below). Adequate quantities of the antigen must be easily obtainable; and if the vaccine is to be used in humans, it is necessary to establish its safety, e.g., it should not induce antibodies which cross-react with human tissue. This is especially important in the development of a streptococcal vaccine, considering the current literature on this subject. However, a recent study has provided evidence that antibodies to *S. mutans* do not cross-react with human heart tissue (Swartzwelder et al. 1988; Russell 1987). In terms of delivery, it is important to ensure that the vaccine will reach the appropriate inductive site in adequate quantities and appropriate form for an effective induction of a mucosal response. This may require enteric coating of encapsulated antigen for delivery of the vaccine to GALT.

Finally, it should be mentioned that other forms of vaccines have received much attention recently and these include anti-idiotypic antibodies, synthetic peptides and carrier organisms expressing cloned microbial gene proteins (reviewed in Roitt 1984).

Oral Adjuvants

Several substances have been shown to be potentially useful oral adjuvants. The adjuvant and targeting effects of cholera toxin have led to recent investigations on the usefulness of this microbial component in oral vaccines (see Elson et al. in this book).

Another widely used adjuvant is synthetic muramyl dipeptide (MDP) which has been extensively characterized by a number of investigators, especially Chedid and his coworkers (1985). A number of investigators, including ourselves (discussed below), have provided evidence to show that MDP, liposomes and recombinant gram-negative bacteria exhibit adjuvant activity when given by the oral route (reviewed by Mestecky 1987). It should be pointed out that the dose used is an important factor to consider for obtaining an adjuvant effect.

USE OF LIPOSOMES IN THE DEVELOPMENT OF ORAL VACCINES

Liposomes have received much attention during the past several years for their usefulness in targeted drug delivery, as adjuvants, and as carriers/vehicles for a variety of substances including antigens and antibodies for the localization/activation of specific cell types (Ram and Tyle 1987; Rouse 1982; Ostro 1987).

Our group has been especially interested in the usefulness of liposomes as vehicles and adjuvants for inducing mucosal responses to *S. mutans* antigens. Liposomes are microscopic closed vesicles composed of a bilayered phospholipid membrane surrounding an aqueous solution. The antigen can be incorporated into the vesicle or inserted into the lipid membrane. Liposomes can be multilamellar or unilamellar and of various sizes depending on the method of preparation. These properties, in addition to the constituents used for their preparation, are of importance in determining how effective liposomes will be as carriers/vehicles of antigens and as adjuvants.

Preparation of Liposomes

In our studies, we have primarily used unilamellar liposomes with an average diameter of 100 nm, consisting of dipalmitoyl phosphatidylcholine, cholesterol, dicetylphosphate and antigen (Childers et al. 1987). In order to reproducibly prepare uniform liposome vaccines, we use a Microfluidizer™ (Microfluidics Corp, Newton, MA). With this apparatus, liposomes are produced in a high pressure chamber, where the aqueous lipid suspension forms a high energy liquid sheet that causes liposome production. By varying the pressure and cycling time, uniformly sized liposomes can be generated which contain the desired amount of antigen (Mayhew et al. 1984).

Characterization of Liposomes

Procedures which have been used to characterize liposome preparations include electron microscopy (reviewed in Pagano and Weinstein 1978) and dynamic light scatter analysis (Ollivon et al. 1986). These procedures, however, require specialized expertise and equipment for the preparation and analysis of samples, and thus, numerous investigators using liposomes in their studies have not characterized their preparation, and only report the method they used for their generation. Since liposome properties could affect their function, we tested whether flow cytometry (FACS) could be used as a convenient and rapid method for characterizing liposomes.

In this study, liposomes were characterized using a transmission electron microscope (TEM) and a FACS. The three liposome preparations used in this study were: a) an initial liposome preparation obtained following sonication and filtration through a 5 µm filter, b) liposomes processed through the Microfluidizer™ for 10 min at 40 lb/sq. in. of inlet pressure, and c) liposomes processed as in b) above, which received an additional 10 min treatment in the Microfluidizer™ at 80 lb/sq. in. of inlet pressure. Polystyrene beads (130 nm; Fluoresbrite Microspheres™, Polysciences, Inc, Warrinton, PA) served as the standard. The mean diameter of particles in each of the four preparations was determined by measuring the diameters of 100 spheres in TEM micrographs using a digitizer interfaced with a computer. These results were compared with the forward laser light scatter (FSC) analysis of the preparations using a FACS.

FACS analysis was used to set the first marker to contain 90% of the standard beads in the submicron region (Table 1). The second marker was set to allow the differentiation of intermediate sized from large, aggregated beads; 8% were of intermediate size, and 2% were of large size. The mean diameter of the beads, determined from the TEM micrograph, was 97 nm. The sonicated liposome preparation A showed heterogeneous vesicles with 54% of the particles being of submicron size, 19% being of intermediate size, and 27% being of large size relative to the size of the standard beads. These particles had a mean diameter of 166 nm. Liposome preparation B primarily contained particles in the submicron size range with a low % being of relatively large size. The mean diameter of liposomes from preparation B was 126 nm. Preparation C was the most homogeneous with 92% of the particles being of submicron size. These results provide evidence that one can routinely use the FACS to determine various properties of liposome preparations, including homogeneity and relative size. Considering the increasing applications of liposomes, it is important to have convenient methods for rapid characterization of preparations, in order to establish the properties of liposomes which are optimal in different systems.

Table 1. Analysis of liposome preparations by FACS and TEM

	FSC analysis (% counts)			Mean diameter
Sample	Submicron	Intermediate	Large	(nm)
Standard Beads	90	8	2	97
Liposomes				
A	54	19	27	166
B	77	16	7	126
C	92	7	1	67

In Vivo Assessment of Oral Vaccines Consisting of Liposomes and Antigen

In our studies with oral vaccines consisting of *S. mutans* antigen, we have employed an experimental rat model (reviewed in Michalek et al. 1986). Briefly, germfree Fischer rats were gastrically intubated with the vaccine for two consecutive days, challenged with the appropriate virulent *S. mutans* organism 5 days later, and subsequently given the oral vaccine at weekly intervals for 3 weeks. One week after the last immunization, saliva and serum samples were collected for assessment of antibody activity by ELISA, the rats were sacrificed, and the mandibles were removed for microbiological analysis and for assessment of caries activity. A brief summary of our results using oral vaccines consisting of *S. mutans* antigens and adjuvants is presented in Table 2.

S. mutans whole cell antigen, when given orally to rats, was effective in inducing salivary IgA responses. The presence of these antibodies correlated with protection against *S. mutans* infection. The addition of MDP to this vaccine resulted in an enhancement of the salivary response and caries immunity. If, on the other hand, rats were given an oral vaccine consisting of purified *S. mutans* serotype CHO, essentially no salivary IgA response or protection against *S. mutans* infection was observed. When the CHO was incorporated into liposomes and used as the oral vaccine, a good salivary IgA response was induced which corresponded with a reduction in the level of plaque, number of *S. mutans* in plaque and a 38% reduction in caries activity. Finally, the incorporation of lipophilic MDP into this oral vaccine resulted in an augmentation of the salivary response and a 75% reduction in caries activity. We are currently studying the induction of mucosal responses in humans given an oral vaccine consisting of CHO in liposomes.

Table 2. Effect of liposomes and MDP in augmenting salivary IgA responses and caries immunity

Oral vaccine	Salivary IgA antibody activity (ELISA units)	Plaque score	*S. mutans* per mandible ($x10^6$)	Caries protection
Whole cell	42.4	13.2	7.8	40%
Whole cell/MDP	89.7	3.5	3.1	80%
CHO	6.3	17.1	15.0	6%
CHO/liposome	28.9	12.3	8.5	38%
CHO/liposome/MDP	42.8	3.7	0.9	75%

USE OF RECOMBINANT DNA TECHNIQUES IN THE DEVELOPMENT OF ORAL VACCINES

During the past several years, much work has been done using recombinant DNA techniques to develop oral vaccines (see chapter by Curtiss et al., and Keren et al. in this book). Our own studies have used avirulent, recombinant *Salmonella typhimurium* mutants expressing cloned *S. mutans* gene products as oral vaccines (Curtiss 1986; Katz et al. 1987; see chapter by Curtiss et al. in this book). In one series of studies, mice were given a single oral administration of viable, recombinant organisms. *Salmonella* were detected by microbiological techniques in the Peyer's patches, but not in any other tissue, on days 4 and 8 after challenge (Fig. 1). Immunohistological analysis of the Peyer's patches revealed that *Salmonella* bind to cells in the dome region, crypts and parafollicular regions. The detection of *Salmonella* in Peyer's patches was followed by the appearance, by day 8, of specific salivary IgA antibodies to the cloned *S. mutans* antigen. This response persisted for approximately 7 weeks. These results provide further evidence for the usefulness of this approach for developing oral vaccines protective against a variety of infectious diseases.

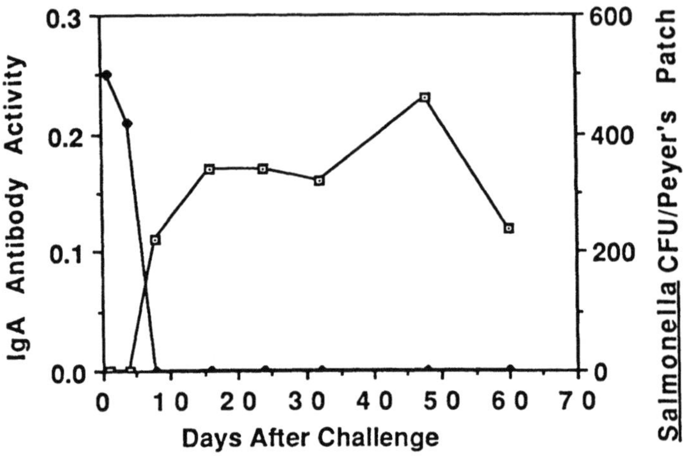

Fig. 1. Salivary IgA antibody activity to the cloned *S. mutans* gene product (closed square) and number of *Salmonella* in Peyer's patches (open square) of mice following oral administration of live recombinant *Salmonella*.

SUMMARY

In this brief review, emphasis was placed on the effectiveness of liposomes as carriers/vehicles of soluble antigens and as adjuvants for mucosal responses when used as oral vaccines. Evidence was provided that oral administration of antigen in liposomes resulted in an augmented mucosal response, compared to the response obtained when the oral vaccine consisted of antigen alone. Specific mucosal responses were further enhanced by the use of lipophilic MDP in the antigen/liposome vaccines. In order to better understand the properties of liposomes important for their functional activities, a rapid and reproducible method employing flow cytometry was described which can be conveniently used for the characterization of liposome preparations. Finally, evidence was presented which further supports the potential of recombinant DNA techniques in developing effective and safe oral vaccines against a variety of infectious diseases.

Acknowledgments. We wish to thank Cecily Harmon and Gloria Richardson for their expert technical help and Ms. Maria Bethune for her effort in preparation of this book. The work from our laboratory which has been reviewed was supported in part by U. S. Public Health Service contract DE 42551 and grants DE 06669, DE 04217, DE 02670 and DE 00155.

REFERENCES

Bockman DE, Cooper MD (1973) Pinocytosis by epithelium associated with lymphoid follicles in the bursa of Fabricius, appendix and Peyer's patches. Am J Anat 136:455-478

Chedid L (1985) Adjuvants of immunity. Ann Inst Pasteur/Immunol 136D:283-291

Childers NK, Michalek SM, Denys F, McGhee JR (1987) Characterization of liposomes for oral vaccines. Adv Exp Med Biol 216B:1771-1780

Curtiss III R (1986) Genetic analysis of *Streptococcus mutans* virulence and prospects for an anticaries vaccine. J Dent Res 65:1034-1045

Czerkinsky C, Prince SJ, Michalek SM, Jackson S, Russell MW, Moldoveanu Z, McGhee JR, Mestecky J (1987) IgA antibody producing cells in peripheral blood after antigen ingestion: Evidence for a common mucosal immune system in humans. Proc Natl Acad Sci (USA) 84:2449-2453

Hamada S, Michalek SM, Kiyono H, Menaker L, McGhee JR (eds) (1986) Molecular microbiology and immunobiology of *Streptococcus mutans*. Elsevier, Amsterdam

Katz J, Michalek SM, Curtiss III R, Harmon C, Richardson G, Mestecky J (1987) Novel oral vaccines: The effectiveness of cloned gene products on inducing secretory immune responses. Adv Exp Med Biol 216B:1741-1747

Mayhew E, Lazo R, Vail WJ, King J, Green AM (1984) Characterization of liposomes prepared using a microemulsifier. Biochim Biophys Acta 775:169-174

McGhee JR, Michalek SM (1981) Immunobiology of dental caries: Microbial aspects and local immunity. Ann Rev Microbiol 35:595-638

Mestecky J (1987) The common mucosal immune system and current strategies for induction of immune responses in external sectrtions. J Clin Immunol 7:265-276

Mestecky J, McGhee JR (1987) Immunoglobulin A (IgA): Molecular and cellular interactions involved in IgA biosynthesis and immune response. Adv Immunol 40:153-245

Michalek SM, Morisaki I, Harmon CC, Hamada S, McGhee JR (1983) Effective immunity to dental caries: Gastric intubation of *Streptococcus mutans* whole cells or cell walls induces protective immunity in gnotobiotic rats. Infect Immun 39:645-654

Michalek SM, Morisaki I, Kimura S, Hamada S, Curtiss III R, McGhee JR (1986) Effective immunity to *Streptococcus mutans*-induced dental caries. In: Hamada S, Michalek SM, Kiyono H, Menaker L, McGhee JR (eds) Molecular microbiology and immunobiology of *Streptococcus mutans*, Elsevier, Amsterdam, p 269

Ollivon M, Walter A, Blumenthal R (1986) Sizing and separation of liposomes, biological vesicles and viruses by high-performance liquid chromatography. Anal Biochem 152:262-274

Ostro MJ (1987) Liposomes. Sci Amer 256:102-111

Owen RL, Jones AL (1974) Epithelial cell specialization within human Peyer's patches: An ultrastructural study of intestinal lymphoid follicles. Gastroenterology 66:189-203

Pagano RE, Weinstein JN (1980) Interactions of liposomes with mammalian cells. Ann Rev Biophys Bioeng 7:435-468

Ram BP, Tyle P (1987) Immunoconjugates: Applications in targeted drug delivery for cancer. Pharm Res 4:181-188

Roitt IM (1984) Immune intervention. Volume 1 New trends in vaccines. Academic Press Inc, London

Rouse BT (1982) Liposomes as carriers of antigens as well as other molecules involved in immunity. J Am Vet Med Assoc 181:988-991

Russell MT (1987) Analysis of heart-reactive antibodies induced in rabbits by immunization with *Streptococcus mutans*. J Oral Pathol 16:234-240

Russell MW, Mestecky J (1986) Potential for immunological intervention against dental caries. J Biol Buccale 14:159-175

Swartzwelder FJ, Barua PK, Albini B, Stinson MW (1988) Heart-reactive antibodies in rabbit anti-*Streptococcus mutans* sera fail to cross-react with *Streptococcus mutans*. J Immunol 140:954-961

Biodegradable Microspheres: Vaccine Delivery System for Oral Immunization

J.H. Eldridge[1], R.M. Gilley[2], J.K. Staas[2], Z. Moldoveanu[1], J.A. Meulbroek[1], and T.R. Tice[2]

[1]Department of Microbiology, University of Alabama at Birmingham, Birmingham, AL 35294; and [2]Controlled Release Division, Southern Research Institute, Birmingham, AL 35255-5305, USA

ABSTRACT

The potential of biocompatible and biodegradable microspheres as a controlled release oral vaccine delivery system has been examined. Orally-administered 1-10 µm microspheres composed of poly (DL-lactide-co-glycolide) were specifically taken up into the Peyer's patch lymphoid tissue of the gut, where those ≥ 5 µm remained for up to 35 days. Microspheres < 5 µm disseminated within macrophages to the mesenteric lymph nodes and spleen. In contrast to soluble staphylococcal enterotoxin B toxoid, oral immunization with enterotoxoid in microspheres induced circulating toxin-specific antibodies and a concurrent secretory IgA anti-toxin response in saliva and gut fluid.

INTRODUCTION

Immunoprophylaxis through vaccination has been responsible for the virtual eradication in the developed countries of several diseases which were once epidemic. Currently, extensive effort is being expended in the identification of appropriate and safe antigens for immunization against numerous infectious agents (Chanock and Lerner 1984). However, the efficiency of these vaccines is in many instances low because of the poor immunogenicity of purified carbohydrates and small proteins. Among the approaches used to enhance vaccine immunogenicity are conjugation to carriers (Arnon et al. 1980) and adjuvants (Warren et al. 1986).

The need for effective vaccination procedures is particularly acute with respect to organisms which produce their pathophysiologic effects through acute infections localized to the gastrointestinal, pulmonary, nasopharangeal, and genitourinary surfaces. These areas are bathed in mucus which contains immunoglobulins consisting almost exclusively of secretory IgA (SIgA) (Hanson 1961; Tomasi and Zigelbaum 1963; Tomasi et al. 1965). This antibody is derived from large numbers of IgA-producing plasma cells which infiltrate the lamina propria regions underlying these mucosal membranes (Brandtzaeg and Baklien 1976; Brandtzaeg 1984), and is specifically transported to the lumenal surface through the action of secretory component (Solari and Kraehenbuhl 1985). However, parenteral immunization regimens are almost universally ineffective at the induction of SIgA responses. Secretory immunity is most often achieved through the direct immunization of mucosally-associated lymphoid tissues, of which the largest mass is represented by the Peyer's patches (PP) of the gastrointestinal tract. Moreover, following their induction in the PP, the precursors of IgA-producing plasma cells extravasate and disseminate to diverse mucosal tissues where final differentiation to high rate IgA synthesis occurs (Crabbe et al. 1969; Bazin et al. 1970; Craig and Cebra 1971). Extensive studies have demonstrated the feasibility of oral immunization to induce this common mucosal immune system (Mestecky et al. 1978), but with rare exception the large doses required to achieve adequate local concentrations in the PP have made this approach impractical for purified vaccine antigens. Among the strategies explored to overcome this problem are incorporation of the antigen into liposomes or the use of genetically engineered bacteria which colonize the PP and express plasmid-encoded vaccine antigens (reviewed by Mestecky 1984). However, practical considerations make both of these approaches unlikely for wide application. Antigen-loaded liposomes are difficult to reproducibly prepare and have a limited shelf life. In the case of a modified bacterial vector the immune response to the carrier microorganism potentially limits its usefulness to a single or widely spaced immunizations. We report here on our studies with biodegradable microspheres consisting of lactide/glycolide copolymers, a vaccine delivery system which results in substantially enhanced immunity without the need for adjuvants and which is effective following oral administration.

MATERIALS AND METHODS

Staphylococcal Enterotoxin B Vaccine

A formalinized vaccine of staphylococcal enterotoxin B was prepared as described by Warren et al. (1978). One gram of enterotoxin was dissolved in 0.1 M sodium phosphate buffer, pH 7.5, to 2 mg/ml. Reagent-grade 36% formaldehyde containing 12% methanol diluted (1:6) in the same phosphate buffer was added to the enterotoxin solution within a length of dialysis tubing to achieve a formaldehyde:enterotoxin molar ratio of 4300:1. The solution was placed in a slowly-shaking 37°C controlled environment incubator-shaker and the pH was checked daily and maintained at 7.5 ± 0.1 units. At the end of the 30 day incubation period the toxoid was concentrated and washed into borate buffered saline (BBS) using a pressure filtration cell (Amicon), and sterilized by filtration. Conversion of the enterotoxin to enterotoxoid was confirmed by the absence of weight loss in 3 to 3.5 kgm rabbits injected intramuscularly with 1 mg of toxoid-converted material.

Microencapsulation

Microspheres containing enterotoxin were prepared by first polymerizing equal molar parts of DL-lactide and glycolide and then purifying the copolymer (DL-PLG). The copolymer was then dissolved so as to incorporate the enterotoxoid into small spherical particles. DL-PLG copolymers are biocompatible esters that biodegrade *in vivo* into lactic and glycolic acids. The mechanism of degradation is by hydrolysis of the ester linkages, and the degradation rate is primarily determined by the ratio of lactide to glycolide present in the copolymer (Miller et al. 1977). Fifty μm microspheres of 50:50 (wt%:wt%) DL-PLG biodegrade to completion in approximately 6 wk after intramuscular injection, and under the same conditions an 85:15 ratio requires > 25 wk.

Immunizations

Groups of BALB/c mice were administered either free or microencapsulated enterotoxoid in 0.5 ml of 8 parts tap water : 2 parts 7.5% sodium bicarbonate via a gastric tube.

Collection of Biologic Fluids

Blood was collected from the retroorbital plexus in calibrated, heparinized capillary pipettes and the plasma harvested following centrifugation. Pools of 100 µl of plasma per mouse from groups of five mice were heat-inactivated and stored at -70°C until assayed. Saliva and gut wash fluids were collected from groups of mice as described by Elson et al. (1984). In brief, the mice were administered four doses (0.5 ml) of lavage solution, isoosmotic with mouse gut fluid, at 15 min intervals via a gastric tube. Fifteen minutes after the last dose the mice were anesthetized and administered 0.1 mg pilocarpine by intraperitoneal injection. The intestinal contents were discharged over a 10 to 20 min period and collected in 5 ml of a solution of 0.1 mg/ml soybean trypsin-inhibitor (Sigma) in 50 mM EDTA. Concurrent with the intestinal discharge, a large volume of saliva is secreted and 200 µl was collected per mouse. The secretion samples were clarified by centrifugation and sodium azide, phenylmethylsulfonyl fluoride, and fetal calf serum were added as preservative, protease inhibitor and alternate substrate for protease activity, respectively. All samples were stored at -70°C until assayed.

Radioimmunometric Assays of Toxin-Specific Antibodies

Radioimmunometric assays were performed in Immulon strips (Dynatech) coated with toxin at 1 µg/ml in BBS, pH 8.4, overnight at 4°C. Control strips were uncoated, but all were blocked for 2 hr at 25°C with 1% bovine serum albumin (Sigma) in BBS, which was also used as the diluent for all samples and ^{125}I-labeled reagents. Various 2-fold dilutions of test samples were added to washed triplicate wells and incubated for 6 hr at 25°C. After washing, 100,000 cpm of ^{125}I-labeled (Hunter 1978) affinity-purified IgG goat anti-mouse µ, α, or γ chain-specific antibody (Southern Biotechnology Associates) was added per well and incubated overnight at 4°C. Following removal of unbound antibodies by washing, the bound ^{125}I-antibodies were detected in a gamma spectrometer (Beckman). The results are presented as the reciprocal of the sample dilution producing a signal > 3-fold that of the group-matched prebleed at the same dilution (end point titration).

Microsphere Tissue Penetration Studies

Mice were administered a single dose of 20 mg of microspheres suspended in 0.5 ml tap water using a blunt-tipped feeding needle inserted into the stomach. The mice were sacrificed at various times and 3 representative PP from the small intestine (one from the midgut and one each from within 1 cm of the stomach and appendix), the first mesenteric lymph node proximal to the appendix, and the spleen were excised, mounted in OCT compound (Miles), and snap frozen in liquid nitrogen. Tissues were cut into 4 to 6 µm serial sections and all sections were viewed in a fluorescence microscope (Leitz) to quantitate microspheres/tissue. The size was determined using a calibrated eyepiece micrometer (Leitz) and its location noted (the PP dome region was defined as the area being within 200 µm of dome epithelium). Some internalized microspheres were divided in the sectioning process, but individual sections contained a limited number of microspheres and multiple viewings of sequential sections allowed divided microspheres to be counted as one and the diameter to be accurately determined.

RESULTS

Several studies have shown that parenteral immunization with antigens trapped within or incorporated into the lipid bilayer of liposomes results in an enhanced antibody response (Allison and Gregoriadis 1974). We investigated the immunopotentiating activity of biocompatible and biodegradable microspheres composed of equal molar parts of polymerized lactide and glycolide (50:50 DL-PLG) on the humoral antibody response to a formalinized toxoid vaccine of staphylococcal enterotoxin B which was homogeneously dispersed within the polymeric matrix. This wall material is from the same class of material currently employed in resorbable sutures. The microspheres examined in these experiments had a size range from 1 to 10 μm and contained 0.23 wt% enterotoxoid.

Based on their probable ability to deliver high local concentrations of antigen and the protection which the wall matrix would be expected to provide against the low pH of the stomach and the proteolytic enzymes of the gut, we speculated that microspheres would represent a suitable vehicle for the delivery of vaccine antigens to the PP. The uptake of microcapsules from the gut lumen by the phagocytic microfold cells (M cells; Bockman and Cooper 1973) which overlie the PP, and their progression into the underlying lymphoreticulum, was investigated by orally-administering a suspension of 85:15 DL-PLG microspheres containing the fluorochrome coumarin-6 to mice. At various times, three representative PP, the spleen, and the first mesenteric lymph node (MLN) draining the appendix were excised, frozen, and serially sectioned. Fluorescence microscopic observation revealed that internalized microspheres of various sizes were present in the PP at 24 hr post administration and at all times tested through 35 d (Table 1). At no time were microspheres of any size observed to penetrate at other sites than the PP. The total number of microspheres observed in the PP increased through 4 d and then decreased over the following 31 d to 15% of the peak number. At 1, 2 and 4 d the proportion of < 2 μm (45-47%), 2 to 5 μm (31-35%) and > 5 μm (18-23%) diameter particles remained constant and reflected the size distribution of the input dose. Concurrent with the decrease in total microsphere numbers beginning at 7 d, a progressive shift in the size distribution was observed such that the small (< 2 μm) and medium (2-5 μm) sizes ceased to predominate and the large (> 5 μm) microspheres became the numerically greatest species observed. In addition, the microspheres observed deep in the lymphoreticulum were exclusively ≤ 5 μm in diameter, and it was this fraction which progressively extravasated from the PP (Table 1).

Table 1. Penetration of coumarin-6 85:15 DL-PLG microspheres into and through the Peyer's patches following oral administration

Time (days)	Total number observed	Proportion of diameter (%) Small <2 μm	Medium 2-5 μm	Large > 5 μm	Proportion at location (%) Dome	Deep
1	296	47	35	18	92	8
2	325	45	32	23	83	17
4	352	46	31	23	76	24
7	196	21	29	41	88	11
14	148	16	29	55	98	2
21	91	7	27	66	98	2
28	63	5	24	71	100	0
35	52	6	19	79	97	3

The peak number of microspheres observed in the MLN occurred at 7 d when their numbers were falling in the PP (Table 2). The size distribution clearly showed that microspheres of > 5 µm in diameter did not enter this tissue. At early time points the majority of the particles in the MLN were located in the subcapsular sinus, but with time they progressed into and through the cortex and medullary regions. Similar examination of the spleens showed that microspheres were not detectable in this organ until 4 d post administration, the peak number occurred at 14 d, and no microspheres > 5 µm in diameter were detected (data not shown). A total of 9 microsphere formulations with different wall materials have been tested and all have been found to enter the PP and migrate through the body with similar kinetics and restrictions imposed by size. In addition, immunohistochemical studies have shown that the < 5 µm microspheres which extravasate the PP do so within phagocytic cells positive for the MAC-1 antigen.

Table 2. Migration of coumarin-6 85:15 DL-PLG microspheres into and through the mesenteric lymph nodes following oral administration

Time (days)	Total number observed	Proportion of diameter (%) Small <2 µm	Medium 2-5 µm	Large > 5 µm	Proportion at location (%) Dome	Deep
1	8	50	50	0	100	0
2	83	76	24	0	95	5
4	97	73	27	0	73	27
7	120	67	32	0	64	36
14	54	83	17	0	9	91
21	20	75	25	0	5	95
28	15	67	32	0	0	100
35	9	44	56	0	0	100

These data suggested that the quality of the immune response induced by orally administering a microencapsulated vaccine could be controlled by the size of the particles. Microspheres < 5 µm in diameter would extravasate the PP within macrophages to stimulate systemic immunity, while those 5 to 10 µm in diameter would stimulate mucosal immunity through the release of antigen in the PP. We addressed these possibilities in experiments in which groups of mice were orally-administered 100 µg of enterotoxoid in solution or as a suspension within 1 to 10 µm 50:50 DL-PLG microspheres, on 3 occasions separated by 30 d. Examination of the plasma endpoint titers of IgM and IgG anti-enterotoxoid antibodies showed that the mice receiving microencapsulated vaccine exhibited a rise in specific antibodies with each oral immunization, while the soluble enterotoxoid was ineffective (Table 3).

Enterotoxin-specific SIgA antibodies were compared in saliva and gut wash samples obtained on days 10 and 20 following tertiary oral immunization with either encapsulated or unencapsulated toxoid (Table 4). In contrast to the inability of the free toxoid to evoke a response when administered orally, the ingestion of an equal amount of the vaccine incorporated into microspheres resulted in a substantial SIgA anti-toxin response in both saliva and gut secretions. It should be pointed out that the gut secretions are diluted into a total of 5 ml per mouse during collection. Although it is impossible to determine the exact dilution factor this imposes, the IgA concentration is at minimum 10-fold higher in the mucus which bathes the gut, and this has not been taken into account in the measurements presented here.

Table 3. Plasma IgM and IgG anti-toxin levels on day 20 following primary, secondary, and tertiary oral immunization with soluble or microencapsulated (50:50 DL-PLG) staphylococcal toxoid

Enterotoxoid dose (μg) per immunization	Form	Plasma anti-toxin titer on day 20 following oral immunization					
		Primary		Secondary		Tertiary	
		IgM	IgG	IgM	IgG	IgM	IgG
100	Microspheres	80	1,280	320	5,120	1,280	40,960
100	Soluble	<20	<20	80	<20	640	<20

Table 4. Toxin-specific IgA antibodies in the saliva and gut fluids of mice on days 10 and 20 after tertiary oral immunization with soluble or microencapsulated enterotoxoid

Enterotoxoid dose (μg) per immunization	Form	IgA anti-enterotoxin titer following tertiary oral immunization			
		Day 10		Day 20	
		Saliva	Gut Wash	Saliva	Gut Wash
100	Microspheres	1,280	1,024	640	256
100	Soluble	40	< 8	10	< 8

DISCUSSION

Results obtained in this study illustrate the potential of biocompatible and biodegradable microspheres as a delivery system for vaccine antigens. Their utility as noninflammatory injectable reservoirs for the uniform or pulsatile long-term release of other bioactive agents is also under active investigation (Redding et al. 1984). Although the immunopotentiating activity of the 1 to 10 μm microspheres seen in these studies may be due in part to a depot effect on antigen release, the lower adjuvancy we have observed following intraperitoneal injection of microspheres of > 10 μm indicates that this is not the only mechanism. It is suggested from the observation that appropriately sized microspheres are rapidly phagocytized by macrophages and that efficient uptake by antigen-presenting accessory cells may play an important role in immune enhancement. In a similar manner, the effectiveness of microspheres as an oral vaccine delivery vehicle is due not only to the protection provided by the wall material during transit through the gut, but also to the efficient targeted delivery of a high concentration of antigen into the PP. In summary, antigen containing microspheres represent a vaccine delivery system with immunopotentiating activity which has the potential for wide application for both parenteral and oral immunizations.

Acknowledgments. The authors are indebted to Drs. Jerry McGhee and Jiri Mestecky for their help and encouragement and to Ms. Sandra Roberts for preparation of this manuscript. This work was supported by USPHS grants AI-24772, AI-21774 and Contract DAMD17-86-C-6162 from the U.S. Army Medical Research Acquisition Activity.

REFERENCES

Allison AC, Gregoriadis G (1974) Liposomes as immunological adjuvants. Nature 252:252

Arnon R, Sela M, Parant M, Chedid L (1980) Antiviral response elicited by a completely synthetic antigen with built in adjuvancy. Proc Natl Acad Sci USA 77:6769-6772

Bazin H, Levi G, Doria G (1970) Predominant contribution of IgA antibody-forming cells to an immune response detected in extraintestinal lymphoid tissues of germ-free mice exposed to antigen by the oral route. J Immunol 105:1049-1051

Bockman DE, Cooper MD (1973) Pinocytosis by epithelium associated with lymphoid follicles in the bursa of Fabricius, appendix and Peyer's patches. An electron microscope study. Am J Anat 136:455-477

Brandtzaeg P (1984) Immune functions of human nasal mucosa and tonsils in health and disease. In: Bienenstock J (ed) Immunology of the lung and upper respiratory tract, McGraw-Hill, New York, p 28

Brandtzaeg P, Baklien K (1976) Immunohistochemical studies of the formation and epithelial transport of immunoglobulins in normal and diseases human intestinal mucosa. Scand J Gastroenterol 11 (Suppl 36):1-45

Chanock RM, Lerner RA (1984) Proceedings of the Cold Spring Harbor Conference: Modern Approaches to Vaccines, Cold Spring Harbor Laboratory

Crabbe PA, Nash DR, Bazin H, Eyssen H, Heremans JF (1969) Antibodies of the IgA type in intestinal plasma cells of germfree mice after oral or parenteral immunization with ferritin. J Exp Med 130:723-744

Craig SW, Cebra JJ (1971) Peyer's patches: An enriched source of precursors for IgA-producing immunocytes in the rabbit. J Exp Med 134:188-200

Elson CO, Ealding W, Lefkowitz J (1984) A lavage technique allowing repeated measurement of IgA in mouse intestinal secretions. J Immunol Meth 67:101-108

Hanson LA (1961) Comparative immunological studies of the immune globulins of human milk and of blood serum. Int Arch Allery Appl Immunol 18:241-267

Mestecky J (1987) The common mucosal immune system and current strategies for induction of immune responses in external secretions. J Clin Immunol 7:265-276

Mestecky J, McGhee JR, Arnold RR, Michalek SM, Prince SJ, Babb JL (1978) Selective induction of an immune response in human external secretions by ingestion of bacterial antigen. J Clin Invest 61:731-737

Miller RA, Brady JM, Cutwright DE (1977) Degradation rates of oral resorbable implants (polylactates and polyglycolates): Rate modification with changes in PLA/PGA copolymer ratios. J Biomed Mater Res 11:711-719

Redding TW, Schally AV, Tice TR, Meyers WE (1984) Long acting delivery systems for peptides: Inhibition of rat prostate tumors by controlled release of D-Trp^6-LH-RH from injectable microcapsules. Proc Natl Acad Sci USA 81:5845-5848

Solari R, Kraehenbuhl J-P (1985) The biosynthesis of secretory component and its role in the transepithelial transport of IgA dimer. Immunol Today 6:17-20

Tomasi TB, Tam EM, Solomon A, Prendergast RA (1965) Characteristics of an immune system common to certain external secretions. J Exp Med 121:101-124

Tomasi TB, Zigelbaum S (1963) The selective occurrence of γ_1A globulin in certain body fluids. J Clin Invest 42:1552-1560
Warren HS, Vogel FR, Chedid LA (1986) Current status of immunological adjuvants. Annu Rev Immunol 4:369-388
Warren JR, Spero L, Metzger JF (1973) Antigenicity of formaldehyde-inactivated staphylococcal enterotoxoid B. J Immunol 111:885-892

Viral Vaccines

Introduction to Oral Immunization Against Viral Infections

R.H. Waldman[1], and K.-C. Bergmann[2]

[1]Department of Internal Medicine, University of Nebraska College of Medicine, Omaha, Nebraska 68105, USA; and [2]Karl-Hansen-Klinik, 4792 Bad Lippspringe, FRG.

INTRODUCTION

There are numerous reasons why oral immunization against a variety of viral infections of mucosal surfaces might become practical for human use. The most important, of course, are immunological ones, i.e., the greater relevance of local mucosal immunity to protection against many pathogens, as compared to systemic immunity. This has been adequately discussed elsewhere in this volume and, therefore, will not be elaborated on here.

PRACTICAL ADVANTAGES

Less frequently discussed, however, are the practical advantages of oral immunization. If one accepts that local immunization has advantages over systemic, then one is faced with the practical difficulties of carrying out such immunization in order to stimulate mucosal immunity. Aside from the gastrointestinal tract, the easiest area to immunize locally is the respiratory tract. Despite the fact that this approach has been successful in certain studies (Waldman et al. 1968; Kasel et al. 1969), it was accomplished only with significant difficulty. Health care personnel are trained to give injections, and significant education must be undertaken to train them to administer aerosol sprays. While many of the general public would prefer such an aerosol spray to an injection, there are many other people who feel quite uncomfortable with having things squirted in their nostrils. In addition, if one has to inhale simultaneously with a dose of aerosol spray, a certain amount of coordination and cooperation is needed. It is unlikely that young children will cooperate sufficiently. If immunizing the respiratory tract is possible, although difficult, immunizing the genital and urinary tracts is impractical or close to impossible. Although studies have been done immunizing intravaginally, this route would not seem to be practical for large scale use (Straus 1961; Waldman et al. 1972).

Oral immunization is convenient. One of the theoretical difficulties with local immunization is that either live, attenuated agents or frequent boosters with killed organisms, must be employed. There are many disadvantages associated with live, attenuated viral immunization, including attaining the proper balance between attenuation and immunogenicity, reversion to virulence, and vaccine liability. Oral immunization permits multiple, even continuous, boosters.

Another factor with respect to convenience is that it might be possible to keep oral vaccines in the home in a lyophilized enterically-coated capsule in the medicine cabinet, or possibly in the refrigerator.

Current Topics in Microbiology and Immunology, Vol. 146

Another advantage of oral immunization is safety. In general, contamination of the oral viral or bacterial vaccine with extraneous material or agents is not a problem. After all, we ingest billions of bacteria each day. This is in contrast to immunization by the parenteral route, or also possibly by the aerosol route. In contrast to parenteral immunization, use of whole organisms for immunization would not seem to be a problem (as it probably is with respect to parenteral immunization with *Bordetella pertussis*). Again, since we ingest myriads of antigens daily, there would seem to be no practical danger from ingesting killed whole organisms.

A final advantage of oral immunization is that it is potentially economical. This would seem to be true even taking into the account the possibility of needing large numbers of organisms. This economy is a result of the fact that physicians' services would probably not be required, i.e., the expense of a doctor's office or clinic visit would be avoided. In addition, as alluded to above, the expense of preparing and safety testing an absolutely pure vaccine product might be avoided.

EARLY STUDIES

Since they are not mentioned elsewhere in this Symposium, we should like to pay brief "homage" to our scientific "ancestors", who in the recent past were pioneers in the development of oral immunization against viral infections. Alexandrova, Zhilova and co-workers, over 15 years ago, carried out a series of studies using a live, attenuated influenza vaccine administered orally. In one study, they administered a bivalent vaccine to children prior to the appearance of a new strain, A/Victoria/35/72 (Zhilova et al. 1977). Serum antibodies were produced to both the vaccine strains, and the new strain (Table 1). Another study showed that the vaccine administered orally induced protection against natural challenge (Table 2).

Table 1. Oral immunization with live attenuated influenza vaccine*

Test strain	% with $\geq$ 4-fold serum antibody rise	Geometric main titer
A/Hong Kong/1/68+	60%	45
A/Victoria/72	30%	17
B/USSR/69+	40%	18

* Children aged 7-15 yrs. (Zhilova et al. 1977)

+ Immunizing strains

Table 2. Protective efficacy of oral live vaccine*

	% with influenza	
Placebo	19%	(1300/9100)
Vaccine	11%	(760/9400)

* Zhilova et al. 1977

Slightly later, Debarbieri and co-workers (1976) carried out a study in 217 subjects, mainly elderly individuals, with an oral live attenuated influenza vaccine. They found that there were no significant side effects from immunization. There was a serum antibody response in 91% of the subjects, and during an epidemic, the vaccine showed a protective efficacy of 73% (Table 3).

Table 3. Protective efficacy of oral live vaccine*

	% with influenza	
Control	12%	(8/66)
Vaccine	3%	(5/151)

* Debarbieri et al. 1976

Unfortunately, secretory antibody measurements were not performed in any of the studies mentioned above.

MOST PROMISING APPLICATIONS

It is clear that respiratory tract infections, particularly with the influenza virus, are a prime target for the practical application of oral immunization. Protection against other respiratory viral infections, even the multitude of strains of rhinovirus, should not be discounted as impossible. The ease with which multiple immunizations can be carried out would seem to make immunization against the most common strains of rhinovirus a possibility.

Immunization against genital tract pathogens would also seem to be a reasonable possibility. Thus, oral immunization using killed herpes simplex viruses should be tested. Finally, protection of newborns by orally immunizing pregnant women so that antibody is stimulated in breast milk could turn out to be quite beneficial.

REFERENCES

Debarbieri A, Crovari P, Cuneo-Crovari P, Badolati G, Giacometti G (1976) Preliminary results of oral influenza vaccination with live attenuated viruses. Develop Biol Standard 33:213-219

Kasel JA, Hume EB, Fulk RV, Togo Y, Huber M, Hornick RB (1969) Antibody responses in nasal secretions and serum of elderly persons following local or parenteral administration of inactivated influenza virus vaccine. J Immunol 102:555-562

Straus ED (1961) Occurrence of antibody in human vaginal mucus. Proc Soc Exp Biol 106:617-621

Waldman RH, Kasel JA, Fulk RV, Mann JJ, Togo Y, Hornick RB, Heiner GT, Dawkins AT (1968) Influenza antibody in human respiratory secretions after subcutaneous or respiratory immunization with inactivated virus. Nature 218:594-595

Waldman RH, Cruz JM, Rowe DS (1972) Immunoglobulin levels and antibody to *Candida albicans* in human cervicovaginal secretions. Clin Exp Immunol 10:427-434

Zhilova GP, Alexandrova GI, Zykov MP, Smorodintsev AA (1977) Some problems of modern influenza prophylaxis with live vaccine. J Infect Dis 135:681-686

Oral Immunization and Secretory Immunity to Viruses

P.L. Ogra[1], E.E. Leibovitz[2], and G. Zhao-Ri[3]

[1]Departments of Pediatrics and Microbiology, State University of New York at Buffalo and Division of Infectious Diseases, Children's Hospital, Buffalo, New York, USA; [2]Pediatrics A, Kaplan Hospital, 76100 Rehovot, Israel; and [3]WHO fellow in the Division of Infectious Diseases, Children's Hospital. Address in China: Beijing Children's Hospital, 42 Nan Li Shi Road, Beijing, People's Republic of China

INTRODUCTION

Available information concerning mucosal immune response to naturally-acquired or vaccine-induced viral infections is largely limited to the development of antibody responses. Limited information exists concerning the appearance and role of virus-specific mucosal T cell-mediated immune responses after oral or systemic immunization in humans. In general, naturally acquired infections, or immunizations by replicating vaccine viruses administered by the mucosal routes which mimic the route of natural infection are associated with the development of effective immune responses in the serum as well as secretory sites, and exhibit protection against mucosal reinfection and development of disease (Piedra and Ogra 1986). In view of the large surface area of the intestinal mucosa and the presence of greater mass of gut-associated lymphoid tissue (GALT), in comparison to the content of lymphoid tissue in the respiratory tract, genital tissue and other mucosal sites, it has been proposed that oral immunization may result in the development of higher magnitude of immune response in the respiratory and other mucosal sites than after immunization by the systemic routes or after direct immunization of other mucosal sites (Mestecky 1987). Recent studies carried out to examine this possibility are briefly summarized here.

HUMAN STUDIES

Cellular Immune Response

Natural Infections and Replicating Viral Vaccines: Studies carried out with naturally acquired varicella-zoster virus (VZV) infection or after immunization with a candidate live attenuated VZV vaccine have shown the induction of *in vitro* proliferative response or development of skin test reactivity to VZV antigens after either form of exposure to the virus. Similarly, the development of cell-mediated immune responses to VZV has been observed in peripheral blood lymphocytes (PBL) as well as tonsillar tissue lymphocytes (TTL). This is exemplified by the observation that during an epidemic exposure to chickenpox, several patients with pre-existing VZV-specific proliferative activity in the TTL were found to be totally protected against disease, when compared to those who lacked such TTL activity (Table 1). It appears that mucosal cellular immunity may be an important determinant of protection against VZV regardless of the corresponding level of cellular immunity in PBL (Bogger-Goren et al. 1984).

Current Topics in Microbiology and Immunology, Vol. 146

Table 1. Effect of pre-existing lymphoproliferative activity (LTF) in the peripheral blood (PBL) and tonsillar lymphocytes (TTL) on the outcome of subsequent household exposure to chickenpox

Antibody status and outcome of VZV exposure	LTF activity prior to exposure to VZV[a]	
	PBL	TTL
Seronegative		
Disease	3.1±1.9	1.7±1.8[b]
No disease	3.6±2.1	7.8±2.2[c]
Seropositive		
Disease	–	–
No disease	3.9±2.4	9.3±3.2[d]

[a] Expressed as mean stimulation index ± SD
[b] Statistical significance: not significant
[c] VS <0.002
[d] VS <0.001

It is not known whether the development of virus-specific mucosal cellular immune response contributes in any detrimental way to the pathogenesis of viral-induced mucosal disease. Studies carried out with respiratory syncytial virus (RSV) employing *in vitro* proliferation of PBL have shown that patients with severe forms of disease (bronchiolitis with wheezing) exhibit significantly higher cellular reactivity to RSV than patients with asymptomatic or mild infections without wheezing or pneumonia (Welliver et al. 1979). As observed with VZV infection, it is possible that TTL reactivity may be better correlated to the course of RSV disease than the PBL, since the disease with RSV is exquisitely restricted to the respiratory tract. Investigations carried out with influenza virus have suggested the induction of virus specific T cells after exposure to different influenza virus proteins (Yamada et al. 1985; Andrews et al. 1987). It has been suggested that T cells mediating delayed hypersensitivity in the absence of specific cytotoxic T lymphocytes may contribute to the development of pulmonary disease (Ada et al. 1981).

Little information is available concerning the development of mucosal cellular immune response to non-replicating viral vaccines in humans or in experimental animal models. However, several recent studies have clearly demonstrated that replication of the virus is not a prerequisite for the development of virus-specific systemic cell-mediated immune responses. Influenza virus-specific cytotoxic T-lymphocyte responses have been elicited with a synthetic peptide of influenza nucleoprotein (Bastin et al. 1987) and recombinant influenza hemagglutinin (Gotch et al. 1987). The implication of these observations are clearly applicable to mucosal cellular immune responses.

Antibody Response

Natural Infection and Replicating Vaccines: Studies carried out with a candidate live attenuated VZV vaccine and naturally acquired VZV infection have shown that replication of virus *per se*, is not an adequate determinant of the mucosal immune response (Baba et al. 1982; Bogger-Goren et al. 1982). Naturally acquired infection, presumably via the respiratory route, was consistently followed by the development of a secretory IgA response in nasopharyngeal secretions, and antibody response to VZV in the peripheral blood. On the other hand, parenteral- or inhalation-induced immunization with the live vaccine failed to induce any secretory IgA response to VZV in the nasopharynx. It should, however, be pointed out that these vaccines regularly induced antibody in the peripheral blood as shown in Table 2 (Bogger-Goren et al. 1984).

Table 2. Development of VZV-specific immunity after natural or vaccine-induced infection

		VZV specific antibody titer after 4 months[a]	
Study group	Virus dose (PFU)	Serum IgG	Secretory IgA
Natural infection	–	1552±900	60±21
VZV vaccine			
S/C[b]	500	44±28	0
I/H	2500	56±15	0
I/H	800	1.3±0.5	0

[a] Expressed as geometric mean ± SD
[b] S/C, subcutaneous; I/H administered via inhalation, PFU plaque forming units of the virus

Immunization with other live viruses, such as RA27/3 attenuated rubella virus vaccine, has been shown to induce a secretory antibody response in respiratory tract after intranasal immunization as well as after subcutaneous immunization. However, subcutaneous immunization with Cendehill or HPV-77 strains of live attenuated rubella vaccine result in minimal or no secretory antibody response in the nasopharynx when administered via either route of immunization (Fishaut et al. 1981; Ogra et al. 1980). Recently, it has been observed that despite the lack of significant nasopharyngeal antibody response following parenteral immunization with HPV or Cendehill strains, such immunization consistently results in the appearance of rubella specific IgA antibodies in milk after immunization in seronegative lactating females. It is suggested that parenteral immunization with rubella vaccine probably permits some transport of viral antigens to the mucosal immunocompetent cell precursors in the respiratory tract, resulting in antigen-induced IgA-B cell activation and possibly their migration to the mammary glands. As a result of viremia associated with parenteral inoculation of the virus, the availability of viral antigens from the blood stream to the breast may induce further proliferation of mucosally-derived immunocompetent cells localized in the breast, and result in development of significant immunologic reactivity in the milk in the absence of such reactivity in the nasopharyngeal secretions (Ogra et al. 1983). Based on the observations summarized above, immunization with live vaccines administered mucosally may thus offer several advantages over immunization via the systemic route.

The development of "alimentary immunity" following immunization with orally administered live attenuated polio vaccine (OPV) is well known (Ogra and Karzon 1971; Ogra et al. 1980). Administration of OPV has been consistently associated with the development of secretory IgA (SIgA) antibody response in the nasopharynx (NPS) as well as intestine (Ogra and Karzon 1971). The level of poliovirus-specific SIgA response has been shown to relate closely to the outcome of subsequent mucosal replication of the virus on a reinfection challenge and high levels of pre-existing antibody have been associated with inhibition of viral replication.

Table 3. Poliovirus type 3 antibody activity in nasopharyngeal secretions after three doses of poliovaccine

	Neutralization		ELISA SIgA	
Vaccine	% positive	Mean titer/ (+SD)	% positive	Mean titer/ (+SD)
OPV OPV OPV (N=6)	100[a]	**10.1** (5.1)	83	**181** (2.1)
IPV IPV IPV (N=15)	40[b]	**4.2** (4.4)	33[c]	**27.9** (5.2)
IPV OPV OPV (N=14)	67	**5.8** (3.9)	93[d]	**48.5** (2.8)
IPV IPV OPV (N=10)	60	**7.0** (5.8)	70	**14.9** (7.0)

a vs b $P<0.05$; c vs d $P<0.01$

Systemic immunization with non-replicating vaccines such as the Salk polio vaccine have been associated with little or no mucosal immune response in the NPS or intestine (Ogra and Karzon 1971). However parenteral priming by Salk vaccine followed by immunization with OPV (Ogra 1984), or repeated immunization with enhanced potency inactivated polio vaccine (IPV) by parenteral route has been associated with the increased development of SIgA activity in NPS (Salk 1980 a,b,c). In a more recent comparative study, approximately 30 to 40% of OPV and 13% of IPV immunized subjects were observed to develop SIgA ELISA, or neutralizing antibody to poliovirus type 3 in the NPS after a single dose of polio immunization. After two vaccine doses, about 60 to 70% of OPV and 20 to 27% of IPV vaccinated subjects exhibited neutralizing or ELISA antibody in the NPS (Zhaori et al. 1988). After three doses of immunization, 100% of OPV and 40% of IPV vaccinees exhibited neutralizing antibody in NPS. It is of interest to note that intramuscular immunization with a single dose of IPV followed by oral immunization with OPV resulted in the development of NPS response in a significantly greater frequency and in higher titers than observed after immunization with three doses of IPV alone (Table 3). In additional studies (Zhaori et al. 1988), the high potency IPV has been found to be highly effective in inducing a serum antibody response which, after three doses of immunization, was often greater in magnitude than after OPV. Both OPV and IPV induced similar serum antibody response against VP1, VP2 and VP3 virion proteins of poliovirus. However, oral immunization with live poliovaccine was found to be more effective in inducing SIgA antibody response in the NPS against VP3 (Table 4). The lack of such response after IPV cannot be explained based on the available data. It is possible that different antigenic determinants of different virion proteins in IPV may be recognized and or accessible to the mucosal immunologic repertoire regardless of the route of immunization. This possibility is supported by the observation that immunization with IPV followed by OPV resulted in more frequent expression of VP3 specific response in NPS than observed after immunization with IPV alone (Table 4).

Table 4. Poliovirus type 3 antibody activity in nasopharyngeal secretions after three doses of poliovaccine

Vaccine	%Subjects positive for SIgA antibody to		
	VPI	VP2	VP3
OPV OPV OPV (N=6)	100	50	67[a]
IPV IPV IPV (N=15)	80	53	7[b]
IPV OPV OPV (N=14)	100	60	53[c]
IPV IPV OPV (N=10)	100	70	30

a vs b P<0.01; b vs c P<0.01

ANIMAL STUDIES

Recently we have initiated studies to determine whether immunization of the GALT *per se* is more effective in inducing an antibody response in the respiratory tract, and other distant mucosal sites (such as genital tract and mammary glands), than observed after direct immunization of the respiratory tract itself.

Development of Antibody Response

Groups of BALB/c mice seronegative for antibody activity against RSV were immunized with ultraviolet light-inactivated or live long strain of RSV administered via intratracheal (I/T), intragastric (I/G), or subcutaneous routes (S/C), with or without *Bordetella pertussis* as an adjuvant. Sham-immunized and uninfected tissue culture immunized animals served as controls. All immunization groups were sampled frequently for the development of RSV-specific antibody response in the serum, NPS, and intestinal contents. The animals were challenged with live RSV administered intranasally 21-28 days after primary immunization. Following primary immunization, varying levels of serum antibody activity were observed in most RSV-immunized animals. The serum IgG response was highest in subcutaneously immunized animals, although appreciable levels were also observed after intragastric immunization. After intratracheal immunization, higher levels of RSV-specific SIgA activity were observed in the nasopharynx when compared to levels observed in intestinal contents. On the other hand, intragastric immunization resulted in higher RSV antibody activity in the intestinal contents than in the nasopharyngeal secretions (Table 5). It is, however, of interest to note that the levels of RSV-specific SIgA activity observed in NPS after intragastric immunization appeared to be similar to those obtained after intratracheal introduction of the virus. These observations suggest that intragastric immunization is equally effective in inducing enteric as well as respiratory tract IgA antibody responses.

Table 5. Antibody response during mucosal immunization with RSV four weeks after immunization

Immunization schedule and route[b]	Number of animals	ELISA RSV antibody titers[a] NPS IgA	Serum IgG	Intestinal IgA
S/C	25	10±2.5	458±1	5±0.5
I/T	18	64±2	38±5	34±2
I/G	28	58±10	108±5	84±2
HEp-2 cell (control)	15	0	0	8±0.6
Sham-immunized (control)	12	0	0	0

[a] Expressed as reciprocal of dilution (mean ± standard error).
[b] S/C subcutaneous, I/T intratracheal, I/G intragastric.

Table 6. Mucosal immunization with RSV and its effect on subsequent respiratory challenge with live RSV

		Virus shedding in the lung after intranasal challenge with live RSV	
Primary immunization route and schedule[b]	Number of animals tested	Mean ± SD virus titer Log_{10}/ml[a]	Lung pathology at challenge
S/C	10	3.2±0.8	++
I/T	15	0.8±0.1	0
I/G	13	2.8±0.5	±
HEp-2	10	3.8±0.6	+
Sham-Immunized	10	3.8±0.6	+

[a] Expressed as Log_{10}/gram of lung suspension/ml. 0-none, + to ++ minimal to moderate viral pathology

[b] S/C subcutaneous, I/G intragastric, I/T intratracheal

Subsequent re-infection challenge with intranasally administered live RSV in RSV-immunized animals resulted in appreciable shedding of virus in the NPS of subcutaneously, or intragastrically immunized animals (Table 6). It should be pointed out that the magnitude of viral shedding in such immunized animals appeared to be lower than observed in the unimmunized or sham-immunized controls. Of significance was the finding that animals immunized with intratracheal administration exhibited the lowest level of virus replication on re-infection challenge. Although pulmonary pathology observed in BALB/c mice after RSV infection was only minimal in most controls and immunized animals, no disease was observed in animals immunized intratracheally (Table 6).

CONCLUSION

The observations summarized in this report have recapitulated recent as well as some earlier observations on the development of mucosal response to certain naturally acquired viral infections and to viral vaccines administered by different routes. Parenteral immunization with certain live (RA27/3 rubella virus) as well as some inactivated vaccines is able to mount varying levels of response in mucosal surfaces. It appears, at least for poliovirus vaccines, that parenteral priming with IPV followed by oral immunization with live poliovaccine is more effective in inducing a higher antibody response in the nasopharyngeal mucosa than repeated parenteral immunization with IPV alone.

Regardless of route of immunization, the levels of virus-specific IgA antibody in the nasopharynx has a major influence in limiting the replication of poliovirus on subsequent reinfection challenge. Animal studies carried out with RSV suggest that intestinal immunization may be as effective as respiratory tract immunization in inducing IgA antibody response in the respiratory tract. However, subsequent reinfection challenge experiments suggest that respiratory immunization may be superior to intestinal immunization, in determining the outcome of subsequent challenge with live virus in the respiratory tract.

REFERENCES

Ada GL, Leung K-N, Ertle H (1981) An analysis of effector T cell generation and function in mice exposed to influenza A or Sendai viruses. Immunol Rev 58:5-24

Andrew ME, Coupar BEH, Boyle DB, Ada GL (1987) The roles of influenza virus haemagglutinin and nucleoprotein in protection: Analysis using vaccinia virus recombinants. Scand J Immunol 25:21-28

Baba K, Yabuuchi H, Takahashi M, Ogra PL (1982) Immunologic and epidemiologic aspects of varicella infection acquired during infancy and early childhood. J Pediatr 100:881

Bastin J, Rothbard J, Davey J, Jones I, Townsend A (1987) Use of synthetic peptides of influenza nucleoprotein to define epitopes recognized by class I-restricted cytotoxic T lymphocytes. J Exp Med 165:1508-1523

Bogger-Goren S, Baba K, Hurley P, Yabuuchi H, Takahashi M, Ogra PL (1982) Antibody response to varicella-zoster virus following natural or vaccine-induced infection. J Infect Dis 146:260-265

Bogger-Goren S, Bernstein JM, Gershon AA, Ogra PL (1984) Mucosal cell-mediated immunity to varicella zoster virus: Role in protection against disease. J Pediatr 105:195-199

Fishaut M, Murphy D, Neifert M, McIntosh K, Ogra PL (1981) Bronchomammary axis in the immune response to respiratory syncytial virus. J Pediat 99:186-191

Gotch F. Rothbard J, Howland K, Townsend A, McMichael A (1987) Cytotoxic T lymphocytes recognize a fragment of influenza virus matrix protein in association with HLA-A2. Nature 326:881-882

Mestecky J (1987) The common mucosal immune system and current strategies for induction of immune responses in external secretions. J Clin Immunol 7:265-276

Ogra PL, Karzon DT (1971) Formation and function of poliovirus antibody in different tissues. Progr Med Virol 13:156-193

Ogra PL, Fishaut M, Welliver RC (1980) Mucosal immunity and immune response to respiratory viruses. In: Weinstein L, Fields B (eds) Seminars in Infectious Diseases, Vol III. Thieme-Stratton New York, p 225

Ogra PL, Losonsky GA, Fishaut M (1983) Colostrum-derived immunity and maternal-neonatal interaction. Ann NY Acad Sci 409:81-95

Ogra PL (1984) Mucosal immune response to poliovaccines in childhood. Rev Infect Dis 6:S361-S368

Piedra PA, Ogra PL (1986) Immunologic aspects of surface infections in the lung. J Pediatr 108:817-823

Salk D (1980a) Eradication of poliomyelitis in the United States. I. Live virus vaccine-associated and wild poliovirus disease. Rev Infect Dis 2:228-242

Salk D (1980b) Eradication of poliomyelitis in the United States. II. Experience with killed poliovirus vaccine. Rev Infect Dis 2:243-257

Salk D (1980c) Eradication of poliomyelitis in the United States. III. Poliovaccines-practical considerations. Rev Infect Dis 2:258-273

Welliver RC, Kaul A, Ogra PL (1979) Cell-mediated immune responses to respiratory syncytial virus infection: Relationship to the development of reactive airway disease. J Pediatr 94:370-375

Yamada A, Ziese MR, Young JF, Yamada YK, Ennis FA (1985) Influenza virus hemagglutinin-specific cytotoxic T cell response induced by polypeptide produced in *Escherichia coli*. J Exp Med 162:663-674
Zhaori G, Sun M, Ogra PL (in press) Characteristics of immune response to poliovirus virion polypeptides after immunization with live or inactivated poliovaccines. J Infect Dis

Oral Immunization with Influenza Virus: Experimental and Clinical Studies

K.-C. Bergmann[1], and R.H. Waldman[2]

[1]Karl-Hansen-Klinik, 4792 Bad Lippspringe, FRG; and [2]Department of Internal Medicine, University of Nebraska Medical Center, Omaha, Nebraska 68105, USA

The idea of a common mucosal immune system with stimulation of intestinal IgA precursor cells which migrate via the blood stream to populate various mucosal surfaces, including the respiratory tract, stimulated new experimental and clinical studies in the area of immunization against influenza.

ANIMAL EXPERIMENTS

Stimulation of Influenza-Specific Antibody

The first evidence for the induction of respiratory tract antibodies against influenza virus following oral immunization with live virus was seen in BALB/c- and CF-1 mice by positive hemagglutination inhibition tests (HIT) in lung lavage fluids. Furthermore, rectally immunized mice (to exclude any possible contamination of the respiratory tract during immunization) showed a significant antibody response (Bergmann and Waldman 1982). In these experiments we have assessed influenza antibody in lavage fluids by a binding assay (HIT and ELISA) (Tischner et al. 1983), and "functional" antibody activity by the plaque-neutralization-test using MDCK cells. BALB/c-mice showed a mean titer (50% plaque reduction) of 1:3.4 in lung lavages (n=5) following two oral immunizations with live vaccine (unpublished results).

Following oral immunization with live A/PR/8/34 virus twice within 14 days, antibody (ELISA) was detected in lung lavage fluids of NMRI-mice 9-11 days after booster immunization (Bergmann et al. 1983b). In the following weeks, the antibody titers increased and reached a maximum 2 or 3 months following the booster immunization. Generally, oral immunization with live influenza vaccines stimulates higher antibody levels in the respiratory tract and serum than does immunization with killed vaccines.

There was no correlation ($p>0.1$) between individual serum and lung lavage fluid antibody levels in orally ($r=0.18$, $n=15$) and subcutaneously ($r=0.06$, $n=20$) immunized mice, supporting the concept of local antibody production in the respiratory tract following oral or parenteral immunizations. Chen et al. (1986) reported that mice orally immunized with inactivated influenza vaccine exhibited IgA and IgG antibodies in lung fluids and sera with a mean lung IgA:IgG ratio of 13:1 when compared with a serum ratio of 0.3:1, further supporting the premise that local IgA responses were induced.

Six intraesophageal immunizations with killed A/PR/8/34 vaccine within 12 days (total of 0.6 mg hemagglutinin and 240 neuraminidase units) stimulated the occurrence, in monkeys (*Macaca rhesus*) of hemagglutination-inhibiting and neuraminidase antibodies in nasal secretions and saliva, without a concomitant rise in

Current Topics in Microbiology and Immunology, Vol. 146

serum antibody (Bergmann et al. 1986a). This study reduced the uncertainty of extrapolation from results in rodents to the human situation and the absence of serum antibody again ruled out the possibility that antibodies in secretions resulted from transudation of antibodies from the circulation.

When BALB/c-mice were immunized twice orally with live vaccine, strain-specific antibodies were detected in homogenates of the urinary bladder, the uterus, and the vagina, in addition to lung lavage fluids (Table 1). There were no correlations between antibody titers in sera and homogenates from urogenital mucosae further suggesting that local antibody production had occurred (Briese et al. 1985). The urogenital antibody secretion found in uterine lavage fluids of NMRI-mice immunized by the same protocol were of the IgA class (Briese et al. 1986).

Table 1. Influenza specific antibody[a] in sera (IgG) and in homogenates of urogenital organs (IgA) after oral immunization of NMRI-mice

Samples	Immunized		Control	
	Number	Titer	Number	Titer
Serum (IgG)	20	640	5	<20
Uterus (IgA)	20	10.0	21	< 1
Urinary bladder (IgA)	20	1.8		n.d.[b]
Vagina (IgA)	20	3.3		n.d.

[a] Expressed as reciprocal mean titer, measured by ELISA.
[b]n.d. = not determined.

In addition to the respiratory and urogenital tracts, influenza antibodies were also found in colostrum and milk, providing suckling offspring with protective antibodies (Bergmann et al. 1983a). Female NMRI-mice were immunized orally with live vaccine twice within 14 days before delivery, and the newborns of immunized and non-immunized mothers were partially exchanged after birth and before the first suckling. Following aerosol virus challenge, infant mice fostered by immunized females showed no mortality and low virus shedding. This was true whether or not they were born to immune females. On the other hand, infant mice born from immune mothers and fostered by non-immune females showed higher virus shedding and some mortality (38%). Infant mice born and suckled by non-immunized mothers showed high virus shedding and 100% mortality (Tischner et al. 1983).

Oral immunization with killed vaccine (Table 2) led to a smaller but still remarkable degree of protection in suckling newborns as demonstrated by a lower virus content and higher survival rate in young mice suckled by immunized mothers. The protective influence of maternally derived serum antibody (placental transfer) was demonstrated by the higher survival rate in newborn mice, born to orally immunized mothers and fostered by control (non-immunized) mothers in comparison to newborns fostered by their own unimmunized mothers.

In colostral samples from orally immunized NMRI-females (live A/PR/8/34 vaccine) antibodies of the IgA (median 1:12) and IgG (median 1:32) class were detected by ELISA (median values of 5 determinations in a pool of 17 samples). The mean IgA level (mg/ml) in colostrum (0.51 $\pm$ 0.07) was similar to the serum level (0.56 $\pm$ 0.1) two days post delivery (unpublished results).

Protection

Oral administration of live and, to a lesser degree, killed influenza vaccine led to protection of young, adult and old mice as assessed by improved survival following viral challenge, and also decreased intensity of respiratory tract infection as assessed by a lower cell yield and lower virus titers in lung lavage fluids (Bergmann et al. 1982; Waldman et al. 1987). Also, the decreased lung compliance due to influenza virus pneumonia can be prevented by oral immunization as has been shown in lung function tests using a multi-chambered body plethysmograph for mice (Bergmann et al. 1984).

Table 2. Protection of newborn mice by milk from mothers orally immunized with killed influenza vaccine

		Lung lavage fluid				
Group suckled by	Number	Cells Number/chamber	Virus[b] Median	Xg	Survival[c] Number	Percent
Immunized mothers	44	716	8	7	15	47
Immunized mothers, born from control mothers	44	764	8	9	16	50
Control mothers, born from immunized mothers	44	816	32	47	9	28
Control mother	44	1056	32	30	1	3

[a] Gamma-inactivated A/PR/8/34 virus was used for oral immunization of mice on days 1, 12 and 26. Pairings were made on days 16-18, deliveries and partial exchange of newborns was done between immunized and unimmunized controls on days 36-39. Aerogenic infection during suckling period was done 16 days post partum, and the lung lavage was performed 3 days later in 12 animals/group.

[b] Reciprocal hemagglutination titer.

[c] 10 days post infection.

Long-Term Effect

The assessment of an immunization program depends not only on the peak protection rate but also on the duration of protection following immunization. Therefore, NMRI-mice were immunized orally with a gamma-inactivated or live vaccine and challenged at different time periods after administration of a lethal virus aerosal (100% mortality within 6 days in non-immunized controls) (Noack et al. 1986). Antibodies in lung lavage fluids were detectable for 6 months and correlated with demonstrable protection. Further, these animals were characterized by a limited number of cells in lung lavage fluids, by low lung virus titers, and most animals survived challenge with virus. The live vaccine led to a greater protection than killed vaccine (Table 3).

Aging

Studies in old (mean age 20 months) female BALB/c-mice showed an age-related decreased secretory antibody response to oral immunization with live virus. In comparison, their rise in serum antibody was much lower and their antibody concentration in lung lavage fluids was only half of that seen in young animals (mean age 3 months, $p<0.05$) (Waldman et al. 1987).

Both old and young mice were protected equally against challenge with live virus when evaluated by the mortality rates, but old mice showed slightly lower protection, as indicated by a more marked increase in the number of lung cells after challenge ($p<0.05$).

CONCLUSIONS FROM ANIMAL EXPERIMENTS

1. Oral administration of killed influenza vaccine stimulates IgA and IgG antibody in respiratory and genital tracts and in colostrum of mice without a concomitant serum antibody rise.

2. Live vaccine induces a greater secretory antibody response and, in addition, a serum IgG antibody response.

3. The occurrence of antibody in the lower respiratory tract and colostrum of mice correlates with protection against aerogenic virus challenge.

CLINICAL STUDIES

Previous studies have used killed (Petrescu et al. 1976) or live (Koval et al. 1979) influenza vaccine given by the oral route to children or to adults in their seventies, however, secretory antibody was not determined in those field studies.

The first evidence for induction of secretory IgA antibody in nasal washings and saliva of volunteers immunized orally with a killed enterically-coated vaccine was published by Waldman et al. (1982a, 1982b, 1986). Oral administration of a liquid vaccine containing 48 µg of HA stimulated no nasal antibody, however, 1 in 5 volunteers did respond when a dose two-fold higher was given. No adverse reactions to the vaccine were noted when several clinical and laboratory tests were performed (Bergmann et al. 1983c).

There was no rise in serum antibody in any of these studies, and there were no side effects related to oral immunization.

Table 3. Long-term results regarding antibody[a] and protection[b] following oral immunization[c] with killed (K) and live (L) influenza vaccine in mice

Time (months)	Antibody in: Serum K	Serum L	Lung lavage K	Lung lavage L	Lung lavage fluid[d]: Virus[e] K	Virus[e] L	Cell yield[f] K	Cell yield[f] L	Mortality(%)[g] K	Mortality(%)[g] L
Preimmune	n.d.[h]	0.0	n.d.	0.0	n.d.	neg.	n.d.	64	–	–
0.5	1.4	3.2	neg.	1.4	8	neg.	1712	82	20	0
2	1.8	3.3	0.3	0.5	4	neg.	504	102	0	0
3	2.3	3.3	0.8	1.4	8	neg.	976	70	40	10
4	n.d.	3.0	n.d.	0.5	n.d.	neg.	n.d.	n.d.	n.d.	0
6	2.5	3.2	0.3	0.5	8	neg.	864	8	10	0
8	1.4	n.d.	neg.	n.d.	4	n.d.	336	20	20	n.d.

[a] Expressed as reciprocal logarithmic mean values of titer (ELISA), in each group of 5 mice.

[b] Following virus aerosol with lethal dose (100% mortality in control mice).

[c] NMRI- mice were immunized twice orally within 10 or 19 days with a total of 20µg HA/mouse with a killed (gamma radiation, 1.6 kGy) or with a live (0.4 ml of 1:10 stock solution ($EID_{50}10^{-7,2}$/0.1 ml, HA titer 1:256) A/PR/8/34 vaccine.

[d] Taken 3 days after virus challenge for each group of 5 mice.

[e] Reciprocal geometric mean of hemagglutination titer, neg. 1= <1:2.

[f] Cell number per chamber.

[g] Each group of 10 mice.

[h] n.d. = not determined.

CONCLUSIONS FROM CLINICAL STUDIES

1. Oral immunization with killed influenza vaccine stimulates IgA antibody in nasal secretions, tears, and saliva without a concomitant rise in serum antibody levels.

2. The orally induced secretory antibody response is dose related.

3. The response is inversely related to the preimmunization titer.

4. The elderly have higher baseline titers in secretions.

5. Children with chronic bronchitis and asthma respond similarly to young healthy adults.

6. Oral influenza vaccines require protection from gastric acid.

7. Oral immunization against influenza seems to be free of any side effects and is easy to perform.

REFERENCES

Bergmann KC, Waldman RH (1982) Occurrence of influenza-specific antibodies in the lung after oral immunization in mice. Poumon Coeur 38:289-292

Bergmann KC, Waldman RH, Veltri, RW (1982) Oral immunization using influenza virus II. Respiratory secretion antibody and protection in mice. Am Rev Resp Dis 125 (Suppl):67

Bergmann KC, Tischner H, Noack K, Pohl W-D (1983a) Maternal-infant transfer of protection against influenza following oral immunization. Proceedings Int Congr Infect Dis, Vienna, 263 P

Bergmann KC, Waldman RH, Noack K, Tischner H, Pohl W-D, Luther P, Stone J (1983b) Oral immunization protects mice against influenza. Fed. Proc. 42:966

Bergmann KC, Muller S, Tischner H, Pohl W-D, Luther P (1983c) Gute Vertraglichkeit bei oraler Immunisierung mit einem Influenza-Adsorbat-Impfstoff. Dt Gesundh-Wesen 38:1904

Bergmann KC, Lachmann B, Noack K (1984) Lung mechanics in orally immunized mice after aerosolized exposure to influenza virus. Respiration 46:218-221

Bergmann KC, Waldman RH, Nordheim W, Nguyen HT, Huynh PL, Tischner H (1986a) Remote-site stimulation of influenza antibodies in monkeys following intestinal stimulation. Immunol Letters 12:65-67

Bergmann KC, Waldman RH, Tischner H, Pohl W-D (1986b) Antibody in tears, saliva and nasal secretions following oral immunization of humans with inactivated influenza virus vaccine. Int Archs Allergy appl Immunol 80:107-109

Bergmann KC, Waldman RH, Reinhofer M, Pohl W-D, Tischner H, Werchan D (1987) Oral immunization against influenza in children with asthma and chronic bronchitis. Adv Exp Med Biol 216B:1685-1690

Briese V, Bergmann KC, Pohl W-D, Noack K, Tischner H, Straube W (1985) Auftreten von Antikörpern gegen Influenzaviren im Urogenitaltrakt und in der Lunge von Mausen nach oraler Immunisierung Klin Med 40:153-155

Briese V, Bergmann KC, Pohl W-D, Tischner H, Straube W (1986) Auftreten von Antikörpern gegen Influenzaviren im weiblichen Genitaltrakt von Mausen nach oraler Immunisierung Zen bl Gynäkol 108:246-250

Chen K-S, Burlington DB, Staton E, Quinnan GV (1986) Preferential induction of hemagglutinin (HA) -specific IgA in the lung of mice orally immunized with killed influenza virus vaccine. Fed Proc 45:692

Jacknowitz A, Howard S, McKenzie R, Chiodo V, Khakoo R, Bergmann KC, Waldman RH (1985) Stability of enteric-coated oral influenza vaccine. 45th Int Congr Pharma Sci, Montreal (FIP -85) abstract 14

Koval TA, Alexandrova GI, Grunis AM, Ivannikov YG, Medvedeva TE, Sychev SV, Shaposhnikova RP, Smorodintsev AA (1979) Effectiveness of peroral and intranasal immunization of school children with live allantoic influenza vaccine - a comparative study. J Hyg Epid Microbiol Immunol 23:194-200

Noack K, Tischner H, Brauniger S, Nordrheim W, Pohl W-D (1986) Langzeitwirkung bei oraler Immunisierung gegen Influenza mit einem Gammastrahleninaktivierten Impfstoff bei Mausen. Z Erkrank Atm-Org 166:286-289

Petrescu A, Mihail A, Popescu A, Cojita M (1976) Bivalent influenza vaccination with inactivated vaccines administered by nasal or oral route. Rev Roum Virol 27:41-45

Stone J, Waldman RH, Khakoo R, Howard S, Jacknowitz A, Charlton J, Johnson S., Bergmann KC (1983) Age-dependent response in humans following oral immunization. Fed Proc 42:967

Tischner H, Noack K, Pohl W-D, Bergmann KC (1983) Nachweis eines immunologischen Schutzes gegen Influenza bei Jungmausen oral immunisierter Muttertiere, Dt Gesundh-Wesen 38:1335-1340

Waldman RH, Lazzell VA, Bergmann KC, Khakoo R, Jacknowitz A, Howard S, Rose C (1982a) Oral immunization against influenza. Poumon-Coeur 38:293-296

Waldman RH, Stone J, Lazzell VA, Bergmann KC, Khakoo R, Jacknowitz A, Howard S, Rose C (1982b) Regulation of the immune response to orally administered antigens. In: Ogra PL, Jacobs DM (eds) Regulation of the immune response, Karger, Basel p 186

Waldman RH, Stone J, Bergmann KC, Khakoo R, Lazzell V, Jacknowitz A, Waldman ER, Howard S (1986) Secretory antibody following oral influenza immunization. Am J Med Sci 29:367-371

Waldman RH, Bergmann KC, Stone J, Howard S, Chiodo V, Jacknowitz A, Waldman ER, Khakoo R (1987) Age-dependent antibody response in mice and humans following oral influenza immunization. J Clin Immun 7:327-332

Immune Responses to Influenza Virus in Orally and Systemically Immunized Mice

Z. Moldoveanu[1], J.K. Staas[2], R.M. Gilley[2], R. Ray[1], R.W. Compans[1], J.H. Eldridge[1], T.R. Tice[2], and J. Mestecky[1]

[1]Department of Microbiology, University of Alabama at Birmingham, Birmingham, AL 35294; and [2]Controlled Release Division, Southern Research Institute, Birmingham, AL 35255-5305, USA

Influenza virus periodically causes epidemics in humans and animals. Therefore, immunization against influenza is of considerable importance. Injection with influenza virus in humans results in the appearance of antibodies to hemagglutinin (HA) and neuraminidase in both serum and nasal secretions (Couch and Kasel 1983; Murphy et al. 1982; Shvartsman and Zykov 1976; Waldman et al. 1968). Although serum antibodies can be correlated with protection against the virus, the presence of secretory antibodies, especially those of the IgA isotype, in secretions of the upper respiratory tract, is thought to be a major determinant of immunity (Bergmann et al. 1982a; Burlington et al. 1983; Clements et al. 1983; Clements and Murphy 1986; Couch and Kasel 1983; Murphy et al. 1981, 1982; Shvartsman and Zykov 1976; Wright et al. 1983).

Although a large volume of data is available concerning immunity to influenza induced by systemic or intranasal immunization (Clements and Murphy 1986; Couch and Kasel 1983; Murphy et al. 1982; Scott and Sydiskis 1976; Shvartsman and Zykov 1976; Waldman et al. 1968), the oral route has been used in a limited number of studies (Bergmann et al. 1982, 1986; Bergman and Waldman 1982, 1983; Boudreault and Pavilanis 1972; Lazzell et al. 1984; Waldman et al. 1987). Oral immunization of mice with inactivated influenza virus induced hemagglutinin-specific secretory IgA (S-IgA) in lung lavages and serum antibodies of IgG isotype, and resulted in protection against challenge with a lethal dose of the virus (Chen et al. 1987).

Influenza vaccine is considered moderately immunogenic, because its efficacy often does not meet the expectations, and because antibodies present in the respiratory tract either lack sufficient specificity or are not present in sufficient quantities. Thus, various adjuvants such as cholera toxin (Chen and Quinnan this volume) have been used to enhance the immunogenicity of a vaccine administered by the oral route. Alternative approaches, such as microencapsulation to protect sensitive bioactive antigens from degradation and control their release over extended periods of time, have been explored (Eldridge et al. this volume).

The goal of our initial studies was to evaluate the feasibility of using microencapsulated preparations of influenza A virus to confer protective immunity when administered to mice by oral or systemic routes. The antigens used for these studies were: monovalent, formalin-inactivated virus vaccine type A, (representative of A/New Jersey /8/76, Hsw1Nsw1, from Merrell National Laboratories Cincinnati, Ohio), and virus type A/Philippines 2/82/X-79 (H_3N_2) (a kind gift from Dr. K.-S. Chen), virulent in mice. The virus, grown in eggs, was purified by sucrose gradient ultracentrifugation, freeze dried, and microencapsulated into the biodegradable and biocompatible polyester, poly-(DL-lactide-co-glycolide). Non-microencapsulated (free) virus was used for comparison.

Current Topics in Microbiology and Immunology, Vol. 146

Mice were immunized with influenza virus by oral or systemic routes using microspheres as a delivery system or, in control groups, as an aqueous suspension of virus at the same protein concentration. For oral vaccination, the antigen was suspended in 0.1M sodium bicarbonate and delivered intragastrically with an animal feeding needle. Systemic immunization was performed intraperitoneally with the microencapsulated or free viral antigen in a saline solution. The specific immune responses were measured in sera and secretions of mice after immunization. Serum was collected before, and at selected time intervals after immunization, with glass capillary tubes from the retroorbital venous plexus of xylazine-ketamine anesthetized mice. Stimulated saliva was collected with capillary tubes after injection with carbamyl choline chloride (1μg/mouse). Tracheal and intestinal lavages were performed according to the method described by Elson et al. (1984).

The collected biological samples were analyzed by ELISA (Murphy et al. 1981) for the presence of specific antibodies against the virus. Polyvinyl microtiter plates were coated with the following antigens: commercially available vaccine type A/swine (H_1N_1), or a suspension of Philippines influenza virus (H_3N_2) grown in MDCK cells. After incubation with samples collected from the immunized mice, bound antibodies were detected by adding biotinylated isotype-specific goat anti-mouse reagents and avidin-peroxidase or peroxidase-labeled goat anti-mouse immunoglobulins, followed by the peroxidase substrate. In selected cases, hemagglutination-inhibition assays were performed.

Mice immunized with influenza type A/Philippines were challenged intranasally with ~5 LD_{50} of virus in 25 μl of saline, while under xylazine-ketamine anesthesia.

IMMUNOGENICITY OF MICROENCAPSULATED VS. FREE VIRUS

In order to determine whether the immunogenicity of the antigen was preserved, we injected a group of mice with identical doses of microencapsulated or free virus. As shown in Table 1, the microencapsulation process did not affect the antigenicity of influenza vaccine. Antibody titers induced by oral or systemic administration of the microencapsulated vaccine were comparable, on day 28 after immunization, with those induced by free virus (Table 1). Despite the lower ELISA antibody titers in sera of mice immunized orally with the microencapsulated material, the hemagglutination-inhibition titers were higher. Subsequent studies using other virus preparations also demonstrated the retention of immunogenicity of the microencapsulated virus.

Furthermore, analyzing the specific antibodies present in sera and secretions of mice orally or systemically immunized with influenza vaccine or whole virus, we observed that the microencapsulated vaccine given by the oral route induced a higher titer of serum antibodies than a similar dose of nonencapsulated antigen (Fig. 1a). Although there was no significant difference in the total level of anti-influenza antibodies (determined by the polyvalent anti-mouse Ig reagent) in the intestinal washes of mice immunized orally with microencapsulated vs. free antigen, the increase in IgA antibodies was more dramatic in mice immunized orally with microencapsulated vs. free antigen (Fig. 1b). A similar pattern was observed for salivary antibodies.

Table 1. Comparison of serum antibody responses induced by influenza vaccine type A ($H_{SW1}N_{SW1}$) in a soluble or microencapsulated form

Route of immunization	Amount of immunogen (µg)	Form of immunization: Solution vaccine – End point titration	HIT	Microencapsulated vaccine – End point titration	HIT
Oral	14	1/3200	1/28	1/1600	1/84
Systemic	2	1/25,000	1/28	1/50,000	1/28
	4	1/50,000	1/84	1/50,000	1/28

HIT - Hemagglutination inhibition titers

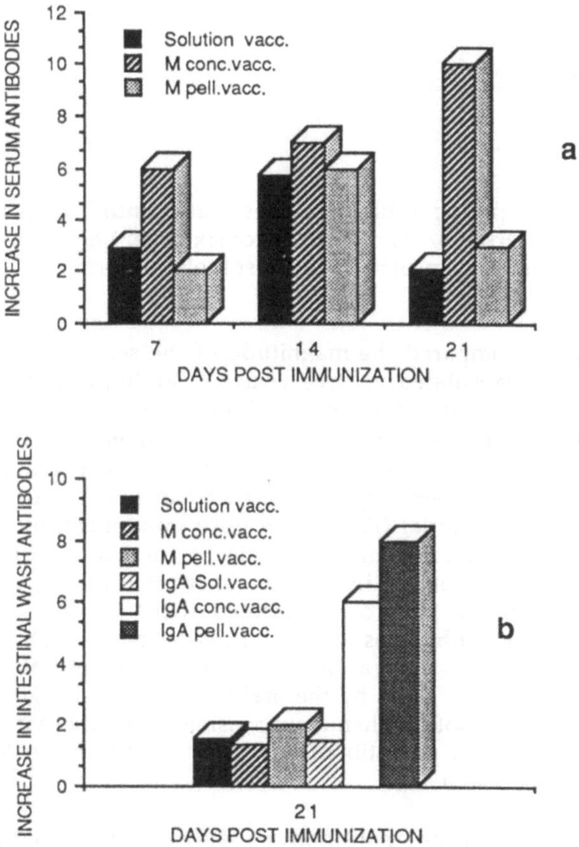

Fig. 1a,b. Increase in serum (a), and intestinal (b) total and IgA-specific antibodies in groups of 6 mice that were orally immunized with influenza virus vaccine (20µg protein/mouse) in a soluble (S) or microencapsulated form (M). Before microencapsulation, the vaccine was either concentrated (conc.) by ultrafiltration, or pelleted (pell.) by ultracentrifugation.

In another group of animals, serum antibodies against influenza virus were monitored for more than four months after immunization. In contrast to mice immunized with free antigen, influenza-specific IgA and IgG antibodies remained elevated in mice immunized with microencapsulated material (Fig. 2).

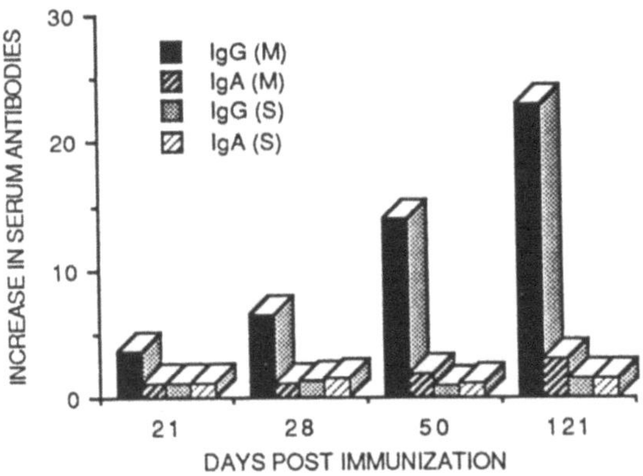

Fig. 2. Increase and duration of specific antibodies after oral immunization of mice with influenza virus (type A/New Jersey 8/76). In groups of six mice, 19µg viral protein was administered per mouse in microencapsulated (M) or soluble (S) form.

In another set of experiments, we compared the magnitude of the secondary immune response induced by the microencapsulated vs. free virus (A/Philippines 2/82/X-79 [H_3N_2]). In systemic immunizations, the microencapsulated antigen induced lower titers of antibodies than the same amount of virus injected in suspension. However, after boosting with same amount of virus, the immune responses in both groups were comparable, or the titers of antibodies were slightly increased in the sera of animals immunized with microencapsulated material (Fig. 3). In mice immunized by the oral route with the same antigen preparations, higher levels of total antibodies were observed in mice primed with solution of influenza virus, but the IgA-specific antibodies were considerably higher in mice orally primed with microencapsulated antigen (Fig. 4). After boosting, total antibody as well as influenza-specific IgA antibody levels were lower in mice that received the antigen in a microencapsulated form. However, the microencapsulated antigen given by the oral route induced higher levels of virus-specific antibodies of the IgA isotype than a suspension of the free virus given by the same route. The increase in specific antibodies in saliva was rather small after systemic or oral immunization (Fig. 5 a,b).

To determine the efficacy of vaccination, non-immune mice, or mice immunized orally or systemically with A/Philippines 2/82/X-79 (H_3N_2) influenza virus either in microspheres or as a suspension, were infected with live virus. Fig. 6, summarizes the results of three challenge experiments. Following systemic immunization with microencapsulated or free virus at doses of 2.5, 5, 10, 20, or 100µg of protein/mouse, all mice survived challenge with $5LD_{50}$ the virus and did not display any sign of disease. Mice immunized orally with of a suspension of free virus manifested signs of illness (such as bristled hair, breathing difficulties, loss of appetite) 2-3 days after a challenge

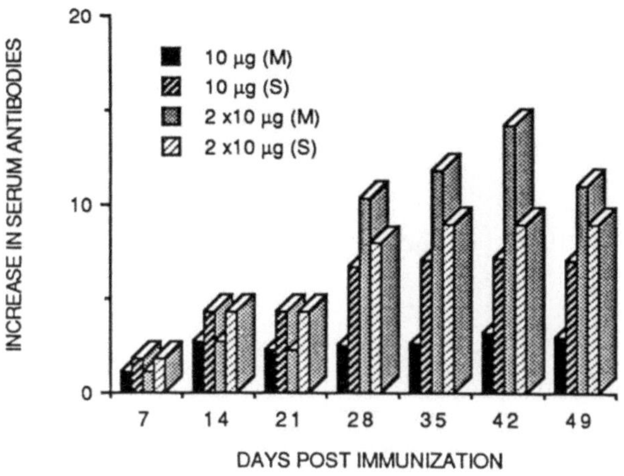

Fig. 3. The increase in specific serum anti-influenza antibodies aftger systemic priming (day 0) and priming followed by boosting (day 21) with the same dose of influenza virus type A/Philippines (µg protein/mouse) in microencapsulated (M) or soluble (S) form.

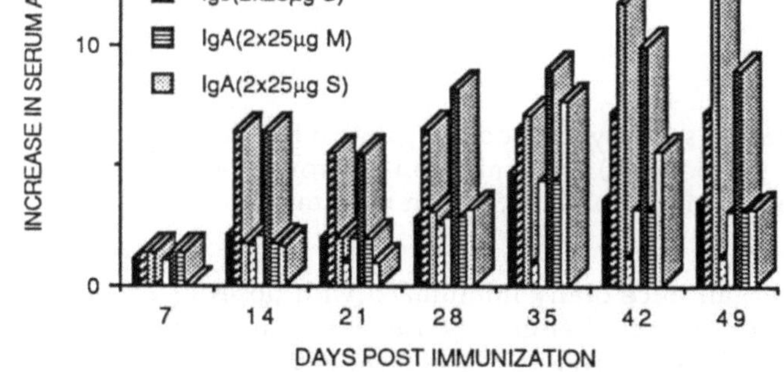

Fig. 4. Serum specific antibody response after oral priming or priming followed by booster immunization with influenza virus type A (H_3N_2) 25µg protein/mouse in microencapsulated (M) or soluble (S) form. Six mice per group were used.

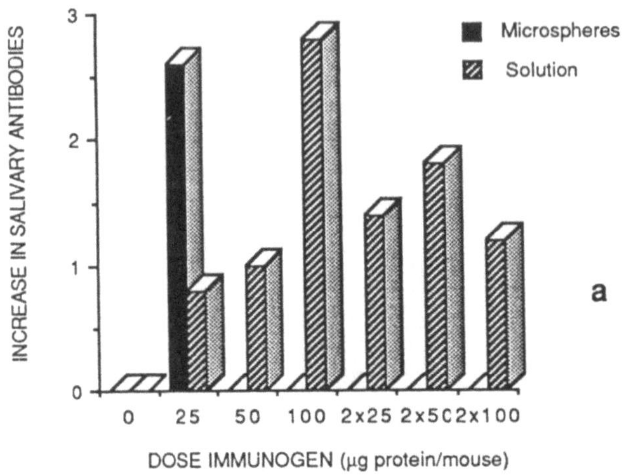

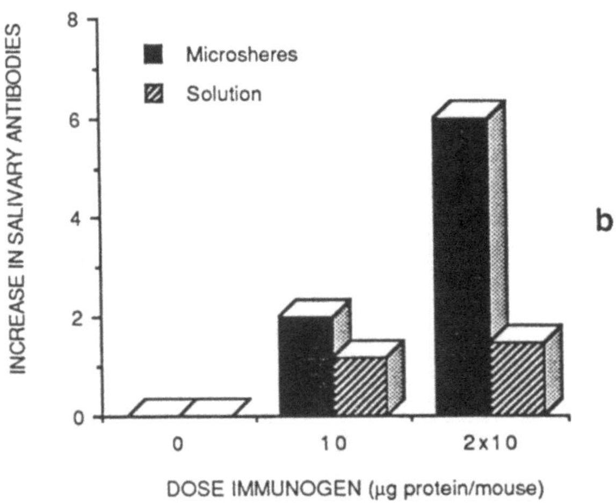

Fig. 5a,b. Increase in specific salivary antibodies against influenza virus type A (H3N2), 49 days after oral (a) or systemic (b) immunization of groups of 6 mice with different doses of microencapsulated (M) or soluble (S) form of immunogen.

with live virus. However, all mice orally immunized with doses of 25, 50, or 100µg of protein/mouse survived the challenge. Oral immunization with 20µg free antigen/mouse resulted in 80% survival. Sixty percent and 33% survival was observed in mice primed and boosted orally with 2.5µg and 5µg of microencapsulated immunogen, respectively. Fifty percent survival of mice only primed with 25µg of microencapsulated material was found, and there was no survival of mice given higher doses of microspheres orally.

Two separate experiments of primary and secondary oral immunization with two different batches of microencapsulated antigen (doses 2.5 - 100μg of protein/mouse) resulted in an unexpected pattern of survival: those mice immunized with high doses of microencapsulated antigen were not protected while those immunized with low doses had a significant survival rate (Fig. 6). In experiments with successive oral followed by systemic immunization or vice versa, there were no survivors in a group of animals primed orally (dose 5, 10, or 20μg of protein/mouse) and boosted systemically with 2.5μg of protein/mouse. In contrast, 50% of mice primed systemically (2.5μg of protein/mouse) and boosted orally (5μg protein/mouse) survived. These preliminary findings suggest that oral immunization with relatively high doses of influenza virus in microencapsulated but *not* in free form induced a state of unresponsiveness to systemic immunization (oral tolerance). Indeed, the levels of serum antibodies were lower in animals immunized with high doses of antigen (Fig. 4 and 5), while those immunized with lower doses had demonstrable levels of serum antibodies.

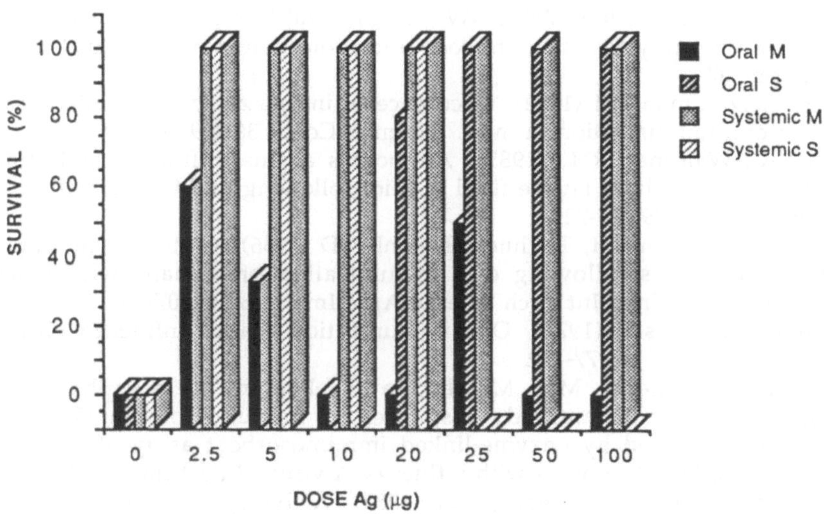

Fig. 6. Percentage of survival after challenge of groups of 6 mice previously orally or systemically immunized with microencapsulated (M) or free (S) antigen. The missing bars correspond to immunizations not performed.

SUMMARY

In our studies on the induction of an immune response by oral immunization, we have explored the potential of a novel approach for antigen delivery by microencapsulation. This procedure preserved the immunogenicity of the influenza virus introduced by either systemic or oral routes. Furthermore, the levels of specific antibodies in serum and in saliva were enhanced and lasted longer (up to 4 months) in animals immunized with of antigens in microencapsulated form than in animals immunized with equal doses of free suspension.

Preliminary challenge experiments showed a correlation between levels of antibodies and protection. All mice systemically immunized were protected against the virus, while mice orally immunized with lower doses of microencapsulated antigen had better survival rates than those immunized with higher doses. Additional experiments suggested that low doses of immunogen were able to generate better protective immunity than high doses, which may instead be tolerogenic.

Further experiments with a well characterized microencapsulated antigen (size of microcapsules, time of release of antigen, as well as its dose and form) will be necessary to establish conditions for optimal immunization protocols applicable for the oral or systemic routes.

Acknowledgments. We wish to thank Ms. Nancy O'Neal for her technical assistance. Supported by USPHS grants AI-18745 and AI-24772.

REFERENCES

Bergmann K-C, Waldman RH, Veltri RW (1982) Oral immunization using influenza virus. II. Respiratory secretion antibody and protection in mice. Am Rev Resp Dis 125(4): Suppl. 67

Bergman K-C, Waldman RH (1982) Occurence of influenza specific antibodies in the lung after oral immunization in mice. Poumon Coeur 38:289-292

Bergmann K-C, Waldman RH (1983) Antibodies against influenza and stimulated macrophages in the lung lavage fluid of mice following oral immunization. Allerg Immunol (Leipzig) 29:215-222

Bergman K-C, Waldman RH, Tischner H, Pohl WD (1986) Antibodies in tears, saliva and nasal secretions following oral immunization of humans with inactivated influenza virus vaccine. Int Arch Allergy Appl Immunol 80:107-109

Boudreault A, Pavilanis V (1972) Oral immunization against influenza virus. Arch Gesamt Virus Forsch 38:177-182

Burlington DB, Clements ML, Meiklejohn G, Phelan M, Murphy BR (1983) Hemagglutinin-specific antibody responses in immunoglobulin G, A, and M istotypes as measured by enzyme-linked immunosorbent assay after primary or secondary infection of humans with influenza A virus. Infect Immun 41:540-545

Chen K-S, Burlington DB, Quinnan Jr GV (1987) Active synthesis of hemagglutinin-specific immunoglobulin A by lung cells of mice that were immunized intragastrically with inactivated influenza virus vaccine. J Virol 61:2150-2154

Clements ML, O'Donnell S, Levine MM, Chanock RM, Murphy BR (1983) Dose response of A/Alaska/6/77/H_3N_2) cold-adapted reassortant vaccine virus in adult volunteers: Role of local antibody in resistance to infection with vaccine virus. Infect Immun 40:1044-1051

Clements ML, Murphy BR (1986) Development and persistence of local and systemic antibody responses in adults given live attenuated or inactivated influenza A virus vaccine. J Clin Microbiol 23:66-72

Couch RB, Kasel JA (1983) Immunity to influenza in man. Annu Rev Microbiol 37:529-549

Elson CO, Ealding W, Lefkowitz J (1984) A lavage technique allowing repeated measurement of IgA antibody in mouse intestinal secretions. J Immunol Meth 67:101-108

Lazzell V, Waldman RH, Rose C, Khakoo R, Jacknowitz A, Howard S (1984) Immunization against influenza in humans using an oral enteric-coated killed virus vaccine. J Biol Stand 12:315-321

Murphy BR, Phelan MA, Nelson DL, Yarchoan R. Tierney EL, Alling DW, Chanock RM (1981) Hemmaglutinin-specific enzyme-linked immunosorbent assay for antibodies to influenza A and B viruses. J Clin Microbiol 13:554-560

Murphy BR, Nelson DL, Wright PF, Tierney EL, Phelan MA, Chanock RM (1982) Secretory and systemic immunological response in children infected with live attenuated influenza A virus vaccines. Infect Immun 36:1102-1108

Scott GH, Sydiskis R J (1976) Response of mice immunized with influenza virus by aerosol and parenteral routes. Infect Immu 13: 696-703

Shvartsman YAS, Zykov MP (1976) Secretory anti-influenza immunity. Adv Immunol 22:291-330

Waldman RH, Kasel JA, Fulk RV, Mann JJ, Togo Y, Hornick RB, Heiner GT, Dawkins Jr AT (1968) Influenza antibody in human respiratory secretions after subcutaneous or respiratory immunization with inactivated virus. Nature (London) 218:594-595

Waldman RH, Bergmann K-C, Stone J, Howard S, Chiodo V, Jacknowitz A, Waldman ER, Khakoo R (1987) Age-dependent antibody response in mice and humans following oral influenza immunization. J Clin Immunol 7:327-332

Wright PF, Murphy BR, Kervina M, Lawrence EM, Phelan MA, Karzon DT (1983) Secretory immunological response after intranasal inactivated influenza A virus vaccinations: Evidence for immunoglobulin A memory. Infect Immun 40:1092-1095

Efficacy of Inactivated Influenza Vaccine Delivered by Oral Administration

K.-S. Chen, and G.V. Quinnan, Jr.

Division of Virology, Center for Biologics Evaluation and Research, Food and Drug Administration, 8800 Rockville Pike, Bethesda, MD 20892, USA

INTRODUCTION

Current immunization against influenza infection employs parenteral injection with inactivated influenza virus vaccines. Although this approach stimulates systemic immunity directed against the protective antigen hemagglutinin (HA) of the virus (Laver and Kilbourne 1966), it only confers a partial protection against infection, at least in some cases. Thus, it is desirable to improve the efficacy of influenza vaccines.

Secretory IgA (SIgA) in respiratory tract was shown to play a critical role in controlling influenza infection (Liew et al. 1984). Direct stimulation of the respiratory system with live (Clements and Murphy 1986) or inactivated influenza virus (Dowie 1973; Liew et al. 1984; Clements et al. 1986) induces local production of secretory IgA. However, this approach to immunization has been incompletely evaluated for safety and efficacy.

Virus-specific antibody was detected in lung lavage fluids of mice that were administered live influenza virus orally (Bergmann et al. 1982, 1983) or in external secretions of human subjects who have ingested an inactivated influenza virus vaccine (Bergmann et al. 1986; Lazzell et al. 1984). We have established a murine model to investigate immunological stimulation of pulmonary IgA following oral inoculation of inactivated influenza virus vaccine.

ORAL IMMUNIZATION LEADS TO LOCAL SYNTHESIS OF ANTIGEN-SPECIFIC ANTIBODIES IN THE LUNGS

We have shown (Chen et al. 1987) that HA-specific IgA and IgG antibodies are produced in lung lavage fluids and sera of adult mice which have been inoculated orally with an inactivated influenza virus vaccine. The predominant immunoglobulin (Ig) isotype in lung lavage fluids was SIgA while that in sera was IgG. We have further shown that the HA-specific IgA was synthesized locally by pulmonary lymphocytes which might have migrated from gut-associated lymphoid tissues (Chen et al. 1987). The immune response pattern induced by oral immunization is similar to that stimulated by nasal infection and is distinct from parenteral immunization. Thus, oral immunization preferentially induces SIgA in a distant mucosal tissue, such as lungs.

PULMONARY SECRETORY IMMUNE RESPONSIVENESS IS AFFECTED BY THE FORM OF ANTIGEN

Currently both whole virion and subunit vaccines of inactivated influenza virus, with similar immunogenicity, are used. When administered orally, whole virus vaccine stimulated significantly higher HA-specific IgA titers in lung lavage fluids than 2 of the

Current Topics in Microbiology and Immunology, Vol. 146

3 commercial subunit vaccines (Table 1). The weaker immunogenicity of detergent-disrupted subunit vaccines may be attributed by the smaller sizes of the virus-derived particles in the vaccines and/or greater sensitivity to protease degradation in the small intestine.

Table 1. Immuogenicity of inactivated influenza whole virus and split virus vaccine administered intragastrically[a]

			HA-antibody titers in lung lavage fluids	
Vaccine	Manufacturer	Number of animals	IgA	IgG
Whole virion	I	8	135 ± 5[b]	10 ± 3
Split	I	9	32 ± 3[c]	9 ± 2
Split	II	10	8 ± 8	4 ± 4
Split	III	9	12 ± 6	8 ± 6

[a] Mice were orally vaccinated with a daily dose containing 40 µg HA of each vaccine for 4 consecutive days. Three weeks later, an identical booster regimen was received and lung lavages were collected 7 days after the last oral dose. HA-specific antibody titers were measured by ELISA and expressed by GMT ± SD.

[b] Significantly higher titers than the other 2 groups which received the split vaccines from manufacturers II and III. $p < 0.05$, Kruskal-Wallis test.

[c] Significantly higher titer than the other group that received the split vaccine from manufacturer II. $p < 0.05$, Kruskal-Wallis test.

PULMONARY SECRETORY IMMUNITY IS REGULATED BY SIZE AND NOT BY FREQUENCY OF ANTIGEN DOSE

A dose-response relationship for induction of SIgA in lung lavage fluids was observed following the oral vaccination with whole virion inacitvated influenza vaccine (Chen et al. 1987). When an immunogen was administered in a single dose or as 4 equally divided daily doses in each immunization regimen, comparable levels of pulmonary SIgA were induced. These results indicate the potential feasibility of a single dose oral immunization.

PERSISTENCE OF PULMONARY ANTIBODY SECRETION

The kinetics of production of HA-specific IgA and IgG antibodies in lung lavage fluids differ. The level of SIgA fell by approximately 2- to 3-fold within 2 weeks following the last oral dose of vaccine. The level of SIgA antibodies was sustained thereafter for at least 3.5 months. In contrast, HA-specific IgG titers were consistently lower than those of IgA and were maintained at a constant level in the 4-month study period (Fig. 1). A similar pattern of IgA persistence was reported in saliva following oral vaccination with *Streptococcus mutans* (Challacombe 1983) and in nasal secretions after intranasal inoculation with inactivated influenza virus (Wright et al. 1983).

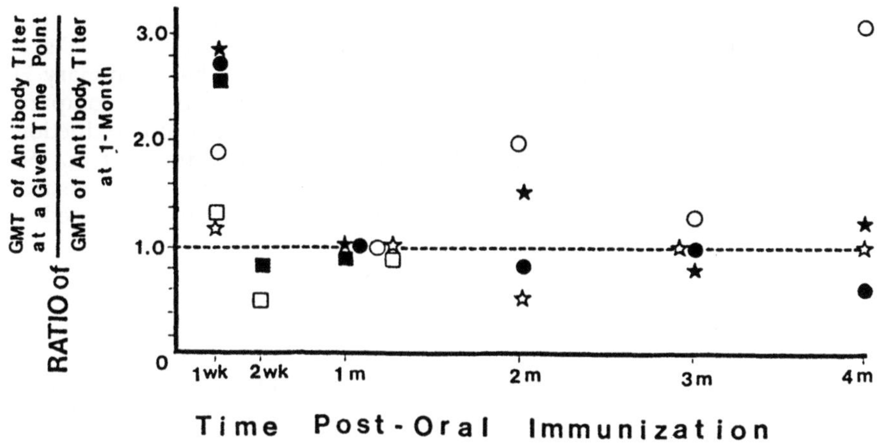

Fig. 1. Persistence of pulmonary HA-specific secretory IgA and IgG antibodies following oral immunization with inactivated influenza virus vaccine. Each mouse received a daily dose containing 40μg HA of inactivated influenza whole virus vaccine for 4 consecutive days in both the primary and booster immunizations. At each time point, after the fourth dose in the booster immunization, lung lavage fluids from 6-8 mice were tested for HA-specific IgA and IgG titers by ELISA. The ratios of geometric mean titers produced by each isotype at various time points to the same isotype secreted at 1-month are expressed. Close symbols (●,★,■) represent pulmonary HA-specific IgA data from individual experiments, while open symbols (○,☆,□) are for pulmonary HA-specific IgG.

CHOLERA TOXIN POTENTIATES PULMONARY ANTIBODY PRODUCTION

Cholera toxin enhances secretion of antigen-specific antibodies in the intestinal tract when it is administered intragastrically concomitantly with an antigen (Elson and Ealding 1984; Lycke and Holmgren 1986). Since the subunit influenza vaccine was weakly immunogenic, we investigated its immune potentiation by cholera toxin.

Our preliminary study demonstrated that simultaneous oral administration of whole virus or subunit influenza virus vaccine with cholera toxin resulted in a significant elevation of HA-specific antibodies in lung lavage fluids and sera compared to administration of vaccines alone. Thus, cholera toxin acts as an adjuvant for secretory immunity at distant mucosal surfaces.

THE CAPACITY FOR SECRETORY IMMUNE RESPONSES IS PRESERVED IN OLDER ANIMALS

The responsiveness of systemic humoral immunity to antigens or mitogens may deteriorate with advancing age (Price and Makinodan 1972a,b; Ershler 1984). Therefore, we examined whether systemic and secretory immunity may be affected by aging. Preliminary results showed that oral immunization of young mice (4 weeks old) resulted in low levels of SIgA antibodies in lung lavage fluids. The pulmonary HA-specific IgA responsiveness matured at 6-7 weeks of age and persisted at a comparable level thereafter (up to 2 years of age). This result is contrary to a 2-fold reduction of pulmonary antibody titers in a study population of 4 mice (Waldman et al. 1987). Pulmonary HA-specific IgG responsiveness was low throughout a span of 2 years. In contrast, levels of HA-specific Ig in sera, of both IgA and IgG isotypes, declined with aging.

CROSS-PROTECTION OF SECRETORY IMMUNITY

In nature, influenza virus constantly undergoes antigenic variations. Effective resistance to influenza infection requires high reactivity of serum neutralizing antibodies with the prevailing strain of influenza virus. SIgA is important in controlling influenza infection (Liew et al. 1984). It may also have greater neutralizing capacity than IgG (Taylor and Dimmock 1985) and may neutralize infective virus *in situ* more efficiently. The data in Table 2 show that mice immunized orally with inactivated influenza A/Philippines/2/82/X-79 (H3N2) vaccine were completely protected from intranasal challenge with a homologous strain of virulent influenza virus. This challenge caused a 90% mortality rate in nonimmunized mice. Those animals immunized with a vaccine comprising a 7 year old isolate, influenza A/England/864/75/Z-49 (H3N2), were protected by 50% against challenge with the heterologous strain of influenza virus, A/Philippines. These studies indicate that secretory immune responses induce a broad cross-protection.

Table 2. Protection of mice after oral immunization against intranasal challenge with virulent influenza virus[a].

Vaccine strain	Mortality after challenge with A/Philippines (H3N2) virus[a] Number dead/Total (%)[b]	
A/Philippines/2/82/X-79 (H3N2)	0/15 (0%)	$p<0.001$
A/England/864/75/X-49 (H3N2)	10/20 (50%)	$p<0.01$
A/Chile/1/83/X-83 (H1N1)	13/17 (76%)	
None	19/21 (90%)	

[a]Each mouse was orally vaccinated for 4 consecutive days with a daily dose of 40 μg HA contained in each whole virus vaccine.
[b]Each mouse was challenged with 5 LD_{50} of mouse-adapted influenza A/Philippines/2/82/X-79 (H3N2) virus.

PERSPECTIVES

Our studies demonstrate that respiratory infection by a viral pathogen can be effectively controlled by oral immunization with an inactivated vaccine. Oral immunization induces a broad cross-protective immunity which is partially effective against distantly related strains of the same subtype of influenza A virus. Furthermore, aged mice retain the capacity to develop secretory immune responses to oral vaccines. Although inactivated vaccines are safe, they require large antigenic doses to be effective. The concomitant oral administration of cholera toxin with inactivated influenza vaccine enhanced secretory immune responses to both whole and subunit virus vaccines. Thus, the identification of a non-toxic adjuvant may allow formulation of an oral vaccine effective at low doses.

The induction of antigen-specific secretory antibodies at distant mucosal surfaces after antigenic stimulation of gut-associated lymphoid tissues may be effective in controlling invasive pathogens.

REFERENCES

Bergmann K-C, Waldman RH (1982) Occurrence of influenza specific antibodies in the lung after oral immunization in mice. Poumon-Coeur 38:289-292

Bergmann K-C, Waldman RH (1983) Antibodies against influenza and stimulated macrophages in the lung lavage fluids of mice following oral immunization. Allergie Immunol 29:215-222

Bergmann K-C, Waldman RH, Tisschner H, Pohl WD (1986) Antibody in tears, saliva and nasal secretions following oral immunization of humans with inactivated influenza virus vaccine. Int Archs Allergy Appl Immunol 80:107-109

Chen K-S, Burlington DB, Quinnan Jr GV (1987) Active synthesis of hemagglutinin-specific immunoglobulin A by lung cells of mice that were immunized intragastrically with inactivated influenza virus vaccine. J Virol 61:2150-2154

Challacombe SJ (1983) Salivary antibodies and systemic tolerance in mice after oral immunization with bacterial antigens. Ann NY Acad Sci 409:177-193

Clements ML, Murphy BR (1986) Development and persistence of local and systemic antibody response in adults given live attenuated or inactivated influenza A virus vaccine. J Clin Microbiol 23:73-76

Downie JC (1973) The sequential appearance of antibody and immunoglobulins in nasal secretion after immunization of volunteers with live and inactivated influenza B virus vaccines. J Hyg 71:433-445

Elson C, Ealding W (1984) Generalized systemic and mucosal immunity in mice after mucosal stimulation with cholera toxin. J Immunol 132:2736-2741

Ershler WB, Moore AL, Socinski MA (1984) Influenza and aging: Age-related changes and the effects of the thymosin on the antibody response to influenza vaccine. J Clin Immunol 4:445-454

Lazzel V, Waldman RH, Rose C, Khakoo R, Jacknowitz A, Howard S (1984) Immunization against influenza in humans using an oral enteric-coated killed virus vaccine. J Biol Standard 12:315-321

Laver WG, Kilbourne ED (1966) Identification in a recombinant influenza virus of structural proteins derived from both parents. Virology 30:493-501

Liew FY, Russell SM, Appleyard G, Brand CM, Beale J (1984) Cross-protection in mice infected with influenza A virus by the respiratory route is correlated with local IgA antibody rather than serum antibody or cytotoxic T cell reactivity. Eur J Immunol 14:350-356

Lycke N, Holmgren J (1986) Strong adjuvant properties of cholera toxin on gut mucosal immune responses to orally presented antigens. Immunology 59:301-308

Price GB, Makinodan T (1972a) Immunological deficiencies in senescence. I. Characterization of intrinsic deficiencies. J Immunol 108:403-412

Price GB, Makinodan T (1972b) Immunological deficiencies in senescence. II. Characterization of extrinsic deficiencies. J Immunol 108:413-417

Taylor HP, Dimmock NJ (1985) Mechanism of neutralization of influenza virus by secretory IgA is different from that of monomeric IgA or IgG. J Exp Med 161:198-209
Waldman RH, Bergmann K-C, Stone J, Howard S, Chiodo V, Jacknowitz A, Waldman ER, Khakoo R (1987) Age-dependent antibody response in mice and humans following oral influenza immunization. J Clin Immunol 7:327-332
Waldman RH, Kasel JA, Fulk RV, Togo Y, Hornick RB, Heiner GG, Dawkins Jr AT, Mann JJ (1968) Influenza antibody in human respiratory secretions after subcutaneous or respiratory immunization with inactivated virus. Nature 218:594-595
Wright PF, Murphy BR, Kervina M, Lawrence EM, Phelan MA, Karzon DT (1983) Secretory immunological response after intranasal inactivated influenza virus vaccinations: Evidence for immunoglobulin A memory. Infect Immun 40:1092-1095

The Systemic and Mucosal Immune Response of Humans to Influenza A Virus

B.R. Murphy[1], and M.L. Clements[2]

[1]Laboratory of Infectious Diseases, National Institute of Allergy and Infectious Diseases, National Institutes of Health, 9000 Rockville Pike, Bethesda, MD 20892; and [2]Center for Immunization Research, Johns Hopkins University School of Hygiene and Public Health, 624 North Broadway, Baltimore, MD 21205, USA

ANTIGENS INVOLVED IN IMMUNITY

Influenza A virus contains a negative-sense, single-stranded RNA genome which consists of 8 segments that code for 7 structural and 3 non-structural proteins (Murphy et al. 1985). In nature, the virus undergoes antigenic changes, referred to as antigenic drift and shift, that permit it to escape from immunity induced by prior infection with related influenza A viruses (Murphy et al. 1985). Common to both types of antigenic change are alterations in the surface glycoproteins, hemagglutinin (HA) and neuraminidase (NA) (Murphy et al. 1985). Antibodies to the HA glycoprotein neutralize the infectivity of the virus, and antibodies to the NA prevent efficient release of virus from infected cells (Askonas et al. 1982; Becht et al. 1971). Passive transfer of monoclonal antibodies to the HA or NA glycoprotein protects mice from experimental challenge with wild type influenza virus whereas those to the membrane (M) protein or nucleoprotein (NP) fail to protect (Askonas et al. 1982). Clearly, the surface glycoproteins constitute the major protective antigens against which an immune response is directed. Immunization of animals with purified NP or with NP expressed by a vaccinia virus induces partial resistance to lethal wild type virus challenge, but the level of resistance is considerably less than that induced by immunization with HA (Andrew et al. 1986; Wraith et al. 1986). It is thought that the resistance induced by immunization with NP is mediated primarily by histocompatibility class I restricted cytotoxic T-cells, but is not limited to this T-cell subset (McDermott et al. 1987; Taylor et al. 1986).

In humans, immunity to the HA, NA, or one of several internal virus proteins each can contribute to resistance to infection or disease caused by influenza A virus (Murphy et al. 1972; Murphy et al. 1973; Couch et al. 1974; McMichael et al. 1983). In the absence of antigenic drift or shift, immunity to influenza virus is long-lived (Noble 1982). This was readily apparent in 1977 when an H1N1 virus reappeared in the human population after it disappearance twenty years previously. Individuals infected with the H1N1 virus in the 1950's remained resistant to disease in 1977. However, this immunity was subtype specific because influenza illness caused by H2N2 and H3N2 viruses was experienced by most of the population during this twenty year interval. Thus, infection with H1N1 influenza viruses bearing highly related internal antigens such as M and NP did not appear to affect the occurrence or consequences of subsequent infections with H2N2 and H3N2 viruses. Since antibody to the HA can neutralize the infectivity of influenza A virus, it is not surprising that immunity induced by this glycoprotein can protect both against infection and disease (Clements et al. 1983; Murphy et al. 1973). In contrast, resistance induced by the NA or internal viral proteins is affected by a reduction in the level and duration of virus replication and by a decrease in the frequency and severity of illness. (Murphy et al. 1972; McMichael et al. 1983). Thus,

Current Topics in Microbiology and Immunology, Vol. 146

antibody to the HA is primarily responsible for preventing infection whereas antibody to the NA and cytotoxic T-cells specific for influenza antigens each decrease virus replication and thereby contribute to recovery from infection.

In the present paper, we will primarily characterize the humoral arm of the immune response following immunization with live or inactivated virus vaccines because of its greater role in resistance to reinfection.

MUCOSAL AND SYSTEMIC IMMUNE RESPONSE OF INFANTS AND CHILDREN TO LIVE OR INACTIVATED INFLUENZA A VIRUS VACCINES

Mucosal Immune Response to Infection

The response to vaccination with live or inactivated virus vaccines is partially dependent on the immune status of the vaccinee resulting from prior infection with influenza A virus. (Wright et al. 1977; Murphy et al. 1987). We will first consider the response to infection with live attenuated influenza A virus vaccine of children who have not previously been infected by influenza A virus (i.e., immunological virgins). Their nasal wash antibody response was characteristic of that expected for a primary viral infection (Johnson et al. 1986; Murphy et al. 1982; Murphy et al. 1987). IgA, IgM and IgG HA antibodies were present in nasal wash specimens within two weeks after immunization. The titer of IgA HA antibody was much higher than that of IgM or IgG. The nasal wash IgA and IgG HA antibody responses persisted for twelve months in about half of the vaccinees. To determine whether the nasal wash IgA, IgG, or IgM HA antibody was actively secreted, the specific activity of antibody (titer of antibody/concentration of Ig) in a nasal wash specimen was compared with that in serum collected at the same time (Murphy et al. 1982). There was evidence for active secretion of each of the three Ig isotypes although this was most frequently observed for IgA. It is expected that IgA and IgM HA antibodies would be actively secreted to a greater extent than IgG HA antibody because of the secretory component-mediated active transport of J chain-containing polymeric immunoglobulin molecules (Brandtzaeg 1981). A local IgA anamnestic antibody response was demonstrated for live virus vaccines (Wright et al. 1983).

The specific activity of nasal wash IgG HA antibody exceeded that in serum in about 20% of the live virus vaccines which suggests that IgG HA antibody might be actively secreted across mucosal epithelium in some children. Additional evidence that supports the hypothesis that IgG antibody can be actively secreted was obtained from a study in children who were immunized parenterally with inactivated virus vaccine and challenged intranasally with live virus vaccine one year later (Johnson et al. 1986). These children had a 25-fold rise in nasal wash IgG HA antibody with only an 8-fold rise in serum IgG HA antibody. Other studies in adults, however, indicated that most of the nasal wash IgG HA antibody appears to be a transudate from serum. (Wagner et al. 1987).

Mucosal Response to Inactivated Virus

The mucosal antibody response of seronegative children to parenteral administration of inactivated virus vaccine differs from that following intranasal administration of live virus vaccine (Johnson et al. 1986; Murphy et al. 1987). The inactivated virus vaccines induced less nasal wash IgG and IgA HA antibodies than did the live virus vaccines, whereas their IgM responses were comparable. IgA and IgM HA antibody titers were virtually absent at six months following immunization with inactivated virus vaccine, whereas antibodies of both isotypes were present in about 50% of the live virus vaccinees one year after immunization.

Serum Antibody Response to Live or Inactivated Virus Vaccine

The serum IgA and IgG HA antibody responses of seronegative children to live virus vaccine were of significantly greater magnitude than those to inactivated virus vaccine (Johnson et al. 1986). The low level of serum antibody response of children and seronegative adults to inactivated virus vaccines represents the basis for requiring two doses of vaccine for these groups whereas one dose is sufficient for seropositive vaccinees (Wright et al. 1977). This pattern of response is interesting mainly because the reciprocal pattern is seen in adults who have had prior experience with influenza A viruses. In such vaccinees, the inactivated virus vaccine is more immunogenic.

We noted an association between the serum and nasal wash IgA antibody response. Individuals infected with virus who developed a nasal wash IgA HA antibody response also developed a serum IgA HA antibody response (Murphy et al. 1982). Furthermore, adults who developed high titers of nasal wash IgA HA antibody also developed high titers of serum IgA HA antibody (Burlington et al. 1983). The possible mucosal origin of this serum IgA antibody will be discussed subsequently.

MUCOSAL AND SYSTEMIC IMMUNE RESPONSES OF ADULTS TO LIVE OR INACTIVATED VIRUS VACCINE

Mucosal Antibody Response

Adult volunteers with little or no detectable serum antibody to influenza A virus HA were administered intranasally a live attenuated virus or an inactivated virus vaccine parenterally in the early 1980's (Clements et al. 1986a). Serum and nasal wash specimens were collected at intervals up to seven months thereafter to assess the immune response to these two immunization procedures. The overall frequency of a nasal wash HA antibody response by the live and inactivated virus vaccines was similar with greater than 90 percent in each group responding but the pattern and magnitude of the responses differed between the two groups. About 95 percent of the inactivated virus vaccinees developed IgG HA nasal wash antibody compared to 60 percent of the live virus vaccinees. The magnitude of the nasal wash IgG response was similar in both groups. About 85 percent of the live virus vaccinees developed an IgA antibody response whereas only 30 to 45 percent of the inactivated virus vaccinees responded. The magnitude of the nasal wash IgA HA antibody response of the live virus vaccinees was 5 to 8-fold greater than that of the inactivated virus vaccinees. The level of the nasal wash IgA HA antibody decreased about 2 to 4-fold over the seven month period whereas the nasal wash IgG titers remained more constant as did IgG HA antibody in serum during this seven month interval.

Serum Antibody Response

The serum antibody responses in the live and inactivated virus vaccinees differed. In general, the magnitude of serum IgA and IgG HA antibody titer rises was about 3 to 4-fold greater in the inactivated virus vaccinees than in those given live virus vaccine.

As mentioned earlier, the greater immunogenicity of the inactivated virus vaccine in adults was not observed in children. The titers of serum IgG antibodies remained stable in adults over a seven month period whereas those in children immunized with inactivated virus vaccine decreased 4-fold over a similar period (Murphy et al. 1987). The level of serum IgA antibodies in both groups of adult vaccinees decreased about 2 to 4-fold over a seven month period.

These results indicated that the response to the live or inactivated virus vaccines is governed in large part by the prior immune status of the vaccinee. Many of the differences between the response of adults and children to live or inactivated virus vaccinees can best be understood in this context. Since replication is partially restricted in adults by immunity induced by prior infection with related influenza A viruses, the live virus vaccine replicates to a higher titer and for a longer period in seronegative children and adults than in seropositive subjects (Wright et al. 1982; Murphy et al. 1980). For this reason, the live virus vaccine presents a much greater antigenic stimulus in the children than in adults. Conversely, the immune response to a non-replicating antigen, i.e., the inactivated virus vaccine is enhanced by prior experience with related antigens. As expected, a greater antibody response was observed in persons with preexisting antigenic experience, i.e., a secondary immune response occurred.

Origin of Nasal Wash IgG Antibody

It is generally thought that immunity at mucosal surfaces is mediated by IgA antibodies. The observations that substantial quantities of IgG antibodies appear in mucosal secretions following infection or parenteral immunization with inactivated virus vaccine was somewhat unexpected, and it deserved further study. The IgG subclass response of the adult vaccinees was subsequently characterized and found to be predominately IgG1 antibody in both serum and nasal wash specimens (Wagner et al. 1987). Several observations from this study suggested that nasal wash IgG HA antibody was derived from the serum by passive diffusion down a concentration gradient (Wagner et al. 1987). First, the ratio of the specific activity of nasal wash to that of serum HA antibody was approximately unity in both live and inactivated virus vaccinees, which is evidence against a significant mucosal IgG HA antibody response. Second, the relative magnitude of IgG subclass antibody titer (IgG1>IgG3>IgG2>IgG4) was similar in nasal washes and in serum. Third, there was a significant correlation between the titers of serum and nasal wash IgG1 HA antibodies. Fourth, the regression lines of the postvaccination IgG serum and nasal wash HA antibody titers for the live and inactivated virus vaccines were almost identical. This suggests that similar mechanisms for the generation of nasal wash IgG HA antibody were operative in the two groups, i.e., passive diffusion of antibody from serum to nasal mucosal surface down a concentration gradient. This hypothesis is offered with the caveat that the findings in two studies suggested that nasal wash IgG antibody were locally generated in some children (Johnson et al. 1986; Murphy et al. 1982). It is reasonable to suggest that the IgG HA antibody that gains access to the mucosal surface comes from the submucosal space by passive diffusion. There can be two sources for this submucosal IgG antibody, serum and submucosal IgG-producing plasma cells. It is possible in children undergoing first infection that the contribution by submucosal IgG-producing plasma cells to this submucosal IgG antibody is substantial and significantly exceeds that derived from serum. Diffusion of this locally produced IgG HA antibody into the luminal space, therefore, could give rise in some children to specific activity of nasal wash IgG HA antibody that exceeds that in serum. The nasal wash IgG in such instances would not in actuality be actively secreted but would be derived by passive diffusion of *locally* synthesized IgG HA antibody.

Origin of IgA HA Specific Antibody in Serum

The origin of IgA HA antibody in the serum was next investigated (Brown et al. 1985; Brown et al. 1987). Since there was a correlation between the magnitude of the serum and nasal wash IgA HA antibody titer in humans infected with influenza A virus, it was suggested that the serum IgA was of mucosal origin (Murphy et al. 1982; Burlington et al. 1983). In individuals infected with wild type or attenuated viruses, it was demonstrated that the serum IgA HA antibody (1) was predominately polymeric, in contrast to total serum IgA which is predominantly monomeric; (2) bound radiolabeled secretory component whereas nasal wash IgA HA antibody did not; and (3) was almost completely IgA1 which was also the predominant subclass in nasal secretions (Brown et al. 1985; Brown et al. 1987). These results suggested that the serum IgA antibodies to HA were of mucosal origin and that influenza A virus HA preferentially stimulated an IgA1 antibody response. When parenterally-administered inactivated virus vaccine was used as the immunogen, the serum IgA HA antibody was also predominantly polymeric. However, more monomeric antibody developed in the inactivated virus than in the live virus vaccinees. The predominance of polymeric IgA following parenteral immunization was unexpected and was seen in both seropositive (H3N2) and seronegative (H1N1) vaccinees. A simple explanation for these findings is not apparent. The polymeric IgA HA antibody response to parenteral immunization of seropositive individuals could be explained if IgA memory cells of mucosal origin seeded peripheral lymphoid tissue and these memory cells were responsible for the majority of the IgA HA antibody response. Clearly, IgA-producing cells are present in the peripheral circulation following infection (Yarchoan et al. 1981). The mechanism of a polymeric IgA HA antibody response of seronegative individuals to parenteral immunization with inactivated virus vaccine is more difficult to understand and is worthy of additional study.

Cell-Mediated Immune Responses

Less is known about the cell-mediated immune response in humans. Live and inactivated virus vaccines can stimulate a cytotoxic T-cell response although the duration of this response lasts less than one year (Ennis et al. 1981; Ennis et al. 1982). The antigens recognized by human cytotoxic T cells have not been completely characterized, but it is clear that individuals differ considerably as to the antigens that they can recognize (McMichael 1986). The generation of cytotoxic T-cell response to primary infection with influenza virus remains uncharacterized.

MECHANISMS OF RESISTANCE TO INFLUENZA VIRUS INFECTIONS IN HUMANS

Both Systemic and Mucosal Immunity Contribute to Resistance

We previously presented evidence that the HA, NA, and internal antigens can be recognized by the cellular or humoral arms of the immune system. In humans it is very difficult to identify the relative contribution of different immune mechanisms to resistance to infection or recovery from infection. Serum antibodies clearly play a role in resistance to influenza A virus in humans (Puck et al. 1980). Infants who possessed high quantities of maternally derived neutralizing antibodies to the influenza A virus HA in their cord blood developed illness due to influenza virus at a later age than did infants with low titers of such antibodies. These passively derived antibodies provided transient but definite protection in the infants. These data are consistent with a wealth of data in animals indicating that passively administered antibodies can provide a high level of protection in the lungs, but a lesser degree of protection in the upper respiratory tract (Askonas et al. 1982; Reuman et al. 1983; Small et al. 1981; Prince et al. 1985).

Infection of humans or animals provides a high level of resistance in both the upper and lower respiratory tract. It is reasonable to suggest that both mucosal and systemic immune responses participate in this resistance. Evidence in humans clearly identifies secretory antibody as a major factor in resistance of the upper respiratory tract to influenza A virus infection or disease (Murphy et al. 1973; Clements et al. 1983; Clements et al. 1986b). The importance of secretory antibody in mucosal immunity to other viruses that infect the mucosal surfaces of the respiratory or gastrointestinal tract of humans has also been demonstrated (Smith et al. 1966; Perkins et al. 1969; Mills et al. 1971).

Efficacy of Live and Inactivated Virus Vaccines

The relative efficacy of live versus inactivated virus vaccines has been studied in humans. Children vaccinated twelve months previously with live or inactivated virus vaccine were challenged with live attenuated virus to determine the effect of immunization on shedding of virus (Johnson et al. 1986). Since the challenge virus was an attenuated virus, the protective efficacy against illness could not be assessed but a comparison of the level of shedding of challenge virus could be made between groups. Children who previously experienced natural infection or who received live virus vaccine demonstrated a dramatic reduction in the amount of virus shed as well as the duration of virus shedding when compared to subjects without prior influenza A experience. In contrast, prior vaccination with inactivated virus vaccine did not reduce the amount or duration of virus shedding when compared to the subjects without prior influenza A experience. These observations made in initially seronegative children indicated that immunity induced by the attenuated virus challenge was greater than that induced by the inactivated virus vaccine. Nasal wash IgA antibody to the influenza HA correlated with this resistance to challenge.

Similar findings in adult volunteers have been observed (Clements et al. 1984), whereas other studies in adults have not revealed a large difference in efficacy between the live and inactivated virus vaccines (Clements et al. 1986c). Live virus vaccinees who received one dose of vaccine exhibited a level of resistance to wild type challenge virus seven months after immunization comparable to that of vaccinees who received two doses of inactivated virus vaccine (Clements et al. 1986c). In each study in which a comparison of live versus inactivated virus vaccine has been undertaken the live virus has induced resistance that was equivalent to or exceeded that induced by inactivated virus vaccine (Clements et al. 1984; Johnson et al. 1986; Clements et al. 1986c).

Antibodies Associated with Resistance

The serum or nasal wash antibody that were associated with resistance to infection or disease caused by experimental challenge of volunteers with wild type influenza A virus were also studied. Protection induced by inactivated virus vaccine against wild type virus infection or illness correlated with the levels of neuraminidase-inhibiting

(NI) and hemagglutination-inhibiting antibody (HAI) in serum as well as the level of nasal wash IgG HA (but not IgA) antibody (Clements et al. 1986a). In contrast, infection-induced resistance correlated with local IgA HA antibody and serum NI antibody titer, but not the HAI antibody titer in serum. These observations suggest that live virus vaccine infection-induced and inactivated virus vaccine-induced immunity may involve different compartments of the immune system. Antibodies in either serum or nasal secretions appear capable of conferring resistance regardless of the isotype of the antibody as long as the antibodies are present in high enough titer to interfere with replication and spread of the virus within the host. Additional evidence for the importance of the humoral arm of the immune response in resistance to reinfection is highlighted by the observations that the regions of the HA molecule that undergo changes during antigenic drift are those regions that bind neutralizing monoclonal antibodies (Wiley et al. 1987).

Cellular Immune Response Associated with Resistance

In humans, only one study has demonstrated an association between the cellular arm of the immune response and resistance to influenza A virus (McMichael et al. 1983). The cytotoxic activity of memory Tcells (class I restricted cytotoxic T cells) correlated with resistance to experimentally administered wild type influenza A virus (McMichael et al. 1983). Volunteers with a high level of cytotoxic T-cell activity shed less virus than those with a low level. The antigens recognized by cytotoxic T cells in humans are highly conserved among influenza A viruses, e.g., the NP and M antigens. However, the epidemiologic behavior of influenza viruses in man suggests that the overall contribution of cytotoxic T cell to reduction of disease during reinfection with influenza A virus is small. This conclusion is deduced from the observation that repeated infection of humans with influenza A virus bearing conserved internal viral antigens provided little resistance to disease caused by new variants of influenza virus, whereas immunity to the homologous virus lasts for over twenty years (Noble 1982). The relative contribution of cytotoxic T cells specific for the different influenza A virus antigens to resistance to influenza A virus in humans remains undefined.

In experimental animals, T-helper or T-cytotoxic clones specific for influenza A virus can protect mice from challenge (Taylor et al. 1986; McDermott et al. 1987). The mechanism by which the transferred cells confer protection and the sites (upper versus lower respiratory tract) at which virus replication is altered requires further study. T cells mediating delayed-type hypersensitivity can potentiate disease following passive transfer, indicating that there is a delicate balance between positive and adverse effects of the cellular arm of the immune response (Leung and Ada 1982).

GOALS OF IMMUNIZATION AND FUTURE DEVELOPMENTS

The goals of immunization against influenza viruses are to prevent the mortality associated with infection in individuals at high risk for complications following influenza virus infection and to protect infants and healthy adults who experience significant morbidity due to influenza virus infection (Glezen 1980). Because the virus will continue to undergo antigenic changes which permit it to escape immunity induced by prior infection or immunization, it is unlikely that we will be able to prevent infection through immunization. Therefore, we will have to be content with preventing the disease caused by influenza A virus infection. New approaches to immunization against influenza A virus include live attenuated virus vaccines, adjuvants to augment antibody responses to inactivated virus vaccines, and immunization of mucosal surfaces with inactivated virus vaccines (Chen et al. 1987; Waldman et al. 1986). Each of the new approaches to immunization needs to be studied systematically for safety and efficacy. The mechanisms by which immunity is induced by each procedure should also be evaluated systematically.

REFERENCES

Andrew ME, Coupar BEH, Ada GL, Boyle DB (1986) Cell-mediated immune responses to influenza virus antigens expressed by vaccinia virus recombinants. Microb Pathogen 1:443-452

Askonas BA, McMichael AJ, Webster RG (1982) The immune response to influenza viruses and the problem of protection against infection. In: Basic and Applied Influenza Research. AS Beare (ed), Boca Raton, Florida, CRC Press p. 159

Becht H, Hammerling U, Rott R (1971) Undisturbed release of influenza virus in the presence of univalent anti-neuraminidase antibodies. Virology 46:337-343

Brandtzaeg P (1981) Transport models for secretory IgA and secretory IgM. Clin Exp Immunol 44:221-232

Brown TA, Clements ML, Murphy BR, Radl J, Haaijman JJ, Mestecky J (1987) Molecular form and subclass distribution of IgA antibodies after immunization with live and inactivated influenza A vaccines. Adv Exp Med Biol 216B:1691-1700

Brown TA, Murphy BR, Radl J, Haaijman JJ, Mestecky J (1985) Subclass distribution and molecular form of immunoglobulin A hemagglutinin antibodies in sera and nasal secretions after experimental secondary infection with influenza A virus in humans. J Clin Microbiol 22:259-264

Burlington DB, Clements ML, Meiklejohn G, Phelan M, Murphy BR (1983) Hemagglutinin-specific antibody responses in immunoglobulin A hemagglutinin antibodies in sera and nasal secretions after experimental secondary infection with influenza A virus in humans. J Clin Microbiol 41:540-545

Chen KS, Burlington DB, Quinnan Jr GV (1987) Active synthesis of hemmagglutinin-specific immunoglobulin A by lung cells of mice that were immunized intragastrically with inactivated influenza virus vaccine. J Virol 61:2150-2154

Clements ML, Betts RF, Murphy BR (1984) Advantage of live attenuated cold-adapted influenza A virus over inactivated vaccine for A/Washington/80 (H3N2) wild-type virus infection. Lancet i:705-708

Clements ML, Murphy BR (1986a) Development and persistance of local and systemic antibody responses in adults given live attenuated or inactivated influenza A vaccine. J Clin Microbiol 23:66-72

Clements ML, Betts RF, Tierney EL, Murphy BR (1986b) Serum and nasal wash antibodies associated with resistance to experimental challenge with influenza A wild-type virus. J Clin Microbiol 24:157-160

Clements ML, O'Donnell S, Levine MM, Chanock RM, Murphy BR (1983) Dose response of A/Alaska/6/77 (H3N2) cold-adapted reassortant vaccine virus in adult volunteers: Role of local antibody in resistance to infection with vaccine virus. Infect Immun 40:1044-1051

Clements ML, Betts RF, Tierney EL, Murphy BR (1986c) Resistance of adults to challenge with influenza A wild-type virus after receiving live or inactivated virus vaccine. J Clin Microbiol 23:73-76

Couch RB, Kasel JA, Gerin JL, Schulman JL, Kilbourne ED (1974) Induction of partial immunity to influenza by a neuraminidase-specific vaccine. J Infect Dis 129:411-420

Ennis FA, Hua QY, Schild GC, Pratt R, Potter CW (1981) HLA-restricted virus-specific cytotoxic T-lymphocyte responses to live and inactivated influenza vaccines. Lancet ii:887-891

Ennis FA, Hua QY, Schild GC (1982) Antibody and cytotoxic T lymphocyte responses of humans to live and inactivated influenza vaccines. J Gen Virol 58:273-281

Glezen WP (1980) Consideration of the risk of influenza in children and indications for prophylaxis. Rev Infect Dis 2:408-420

Johnson PR, Feldman S, Thompson JM, Mahoney JD, Wright PF (1986) Immunity to influenza A virus infection in young children: A Comparison of natural infection, live cold-adapted vaccine, and inactivated vaccine. J Infect Dis 154:121-127

Leung KN, Ada GL (1980) Cells mediating delayed-type hypersensitivity in the lungs of mice infected with an influenza A virus. Scand J Immunol 12:393-400

McMichael AJ, Gotch FM, Noble GR, Beare PAS (1983) Cytotoxic T-cell immunity to influenza. N Engl J Med 309:13-17

McDermott MR, Lukacher AE, Braciale VL, Braciale TJ, Bienenstock J (1987) Characterization and in vivo distribution of influenza-virus-specific T-lymphocytes in the murine respiratory tract. Am Rev Resp Dis 135:245-249

Mills J, Van Kirk JE, Wright PF, Chanock RM (1971) Experimental respiratory syncytial virus infection of adults. J Immunol 107:123-130

Murphy BR, Chalhub EG, Nusinoff SR, Kase J, Chanock RM (1973) Temperature-sensitive mutants of influenza virus III. Further characterization of the ts-1[E] influenza A recombinant. J Infect Dis 128:479-487

Murphy BR, Clements ML, Johnson, PR, Wright PF (In Press) The mucosal and systemic immune responses of children and adults to live and inactivated influenza A virus vaccine. In: Strober W, Lamm M, McGhee J (eds) Mucosal Immunity and Infection at Mucosal Surfaces. Oxford University Press

Murphy BR, Kasel JA, Chanock RM (1972) Association of serum anti-neuraminidase antibody with resistance to influenza in man. N Engl J Med 286:1329-1332

Murphy BR, Nelson DL, Wright PF, Tierney EL, Phelan MA, Chanock RM (1982) Secretory and systemic immunological response in children infected with live attenuated influenza A virus vaccines. Infect Immun 36:1102-1108

Murphy BR, Rennels MB, Douglas Jr RG, Betts RF, Couch RB, Cate Jr TR, Chanock RM, Kendal AP, Maassab HF, Suwanagool S, Sotman SB, Cisneros LA, Anthony WC, Nalin DR, Levine MM (1980) Evaluation of influenza A/Hong Kong/123/77 (H1N1) ts-1A2 and cold-adapted recombinant viruses in seronegative adult volunteers. Infect Immun 29:348-355

Murphy BR, Webster RG (1985) Influenza Viruses. In: Fields B, Knipe DM, Melnick JL, Chanock RM, Roizman B, Shope RE (eds) Virolgy. Raven Press, New York p11

Noble GR (1982) Epidemiological and clinical aspects of influenza. In: Beare AS (ed) Basic and Applied Influenza Research. Boca Raton, Florida, CRC Press, p 1179

Perkins JC, Tucker DN, Knopf HLS, Wenzael RP, Kapikian AZ, Chanock RM (1969) Comparison of protective effect of neutralizing antibody in serum and nasal secretions in experimental rhinovirus type 13 illness. Am J Epidemiol 90:519-526

Prince GA, Horswood RL, Chanock RM (1985) Quantitative aspects of passive immunity to respiratory syncytial virus infection in infant cotton rats. J Virol 55:517-520

Puck JM, Glezen WP, Frank AL, Six HR (1980) Protection of infants from infection with influenza A virus by transplacentally acquired antibody. J Infect Dis 142:844-849

Reuman PD, Paganini CMA, Ayoub EM, Small Jr PA (1983) Maternal-infant transfer of influenza-specific immunity in the mouse. J Immunol 130:932-936

Small Jr PA, Ramphal R, Reuman P, Kris R (1981) Host defense against influenza virus. In: Nayak DP (ed) ICN-UCLA Symposia on Molecular and Cellular Biology. Academic Press, New York, p 441

Smith CB, Purcell RH, Bellanti JA, Chanock RM (1966) Protective effect of antibody to parainfluenza type 1 virus. N Engl J Med 21:1145-1152

Taylor PM, Askonas BA (1986) Influenza nucleoprotein-specific cytotoxic T-cell clones are protective in vivo. Immunology 58:417-420

Wagner DK, Clements ML, Reimer CB, Snyder M, Nelson DL, Murphy BR (1987) Analysis of immunoglobulin G antibody responses after administration of live and inactivated influenza A vaccine indicates that nasal wash immunoglobulin G is a transudate from serum. J Clin Microbiol 25:559-562

Waldman RH, Stone J, Bergmann CH, Khakoo R, Lazzell V, Jacknowitz A, Waldman ER, Howard S (1986) Secretory antibody following oral influenza immunization. Am J Med Sci 292:367-371

Wang ML, Skehel JJ, Wiley DC (1986) Comparative analyses of the specificities of anti-influenza hemagglutinin antibodies in human sera. J Virol 57:124-128

Wraith DC, Vessey AE, Askonas BA (1987) Purified influenza virus nucleoprotein protects mice from lethal infection. J Gen Virol 68:433-440

Wright PF, Murphy BR, Kervina M, Lawrence EM, Phelan MA, Karzon DT (1983) Secretory immunological response after intranasal inactivated influenza A virus vaccinations: Evidence for immunoglobulin A memory. Infect Immun 40:1092-1095

Wright PF, Okabe N, McKee Jr KT, Maassab HF, Karzon DT (1982) Cold-adapted recombinant influenza A virus vaccines in seronegative young children. J Infect Dis 146:71-79

Wright PF, Thompson J, Vaughn WK, Folland DS, Sell SHW, Karzon DT (1977) Trials of influenza A/New Jersey/76 virus vaccine in normal children: An overview of age-related antigenicity and reactogenicity. J Infect Dis 136:731-741

Wiley DC, Skehel JJ (1987) The structure and function of the hemagglutinin membrane glycoprotein of influenza virus. Ann Rev Biochem 56:365-394

Yarchoan R, Murphy BR, Strober W, Clements ML, Nelson DL (1981) *In vitro* production of anti-influenza virus antibody after intranasal inoculation with cold-adapted influenza virus. J Immunol 127:1958-1963

Oral Immunization with Sendai Virus in Mice

J.G. Nedrud, X.-P. Liang, and M.E. Lamm

Institute of Pathology, Case Western Reserve University, Cleveland, Ohio, USA

We have sought to exploit the common mucosal immune system in order to achieve effective respiratory immunity following oral, i.e., gastrointestinal, immunization. The concept of a common mucosal immune system is embodied in the observation that IgA B lymphocytes initially encountering antigen at one mucous membrane can, during their differentiation toward IgA-secreting plasma cells, migrate to other mucous membranes distant from the site of initial antigenic stimulation (Rudzik et al. 1975; Weisz-Carrington et al. 1979). Although antigen is not required for this homing of IgA plasma cell precursors to distant mucous membranes, it has been shown, for example, that antigen applied to the respiratory tract can enhance the number of IgA-bearing, antigen-specific lymphocytes which accumulate in the respiratory tract of orally or enterically primed rodents (Pierce and Cray 1981; Weisz-Carrington et al. 1987). The model system we are using is Sendai virus infection of mice. Sendai virus, a parainfluenza type 1 virus related to human respiratory pathogens, naturally infects mice and can cause a wide spectrum of disease ranging from relatively mild upper respiratory infection to lethal viral pneumonia.

Our approach has been systematic: induction of a vigorous mucosal immune response in the gut was the first concern, examination of immunity in the respiratory tract followed. For determination of respiratory immunity, the upper and lower respiratory tracts were examined separately so that the relative contribution of anti-viral IgA and other classes of antibody could be evaluated for each site. This was accomplished by collecting from individual mice both bronchoalveolar lavage (lower respiratory tract) and nasal washings (upper respiratory tract), which were then evaluated for virus-specific IgA and IgG and, where appropriate, content of virus as well.

Initial experiments indicated that even multiple oral immunizations with relatively large doses (0.5 mg viral protein) of live or inactivated virus elicited only modest anti-viral antibody titers in intestinal contents. Elson and Ealding (1984) as well as Lycke and Holmgren (1986) had reported that cholera toxin possesses adjuvant properties for protein antigens when coadministered to the gastrointestinal tract. Therefore, several cholera toxin preparations were evaluated for their capacity to enhance intestinal anti-viral antibodies. Table 1 shows that when 10 µg of cholera toxin were mixed and administered intragastrically with each of four 0.5 mg doses of inactivated Sendai virus, gut anti-viral IgA titers were increased 10-fold ($p<0.001$). Enhancement of gut IgG anti-viral titers was also significant ($p=.007$). A similar quantity of cholera toxin was also covalently linked to Sendai virus by a two-step glutaraldehyde conjugation process (McKenzie and Halsey 1984) prior to oral immunization. Conjugation of toxin to virus in this manner could theoretically facilitate cell interactions favoring a vigorous mucosal immune response in the gut. In addition, McKenzie and Halsey (1984) have reported that cholera toxin B subunit covalently linked to horseradish peroxidase enhanced gut IgA and IgG responses to horseradish peroxidase. They employed an approximate 1:1 ratio of horseradish peroxidase to cholera toxin B subunit; the virus to cholera toxin mass ratio in our conjugate was 38:1. Oral immunization with this virus-

Current Topics in Microbiology and Immunology, Vol. 146

toxin conjugate resulted in a further three-fold enhancement of gut IgA titers (p=.024) but no significant increase in gut IgG titers (p=0.2) compared to the mixture of virus and toxin (Table 1). Serum IgA and occasionally serum IgG titers were also increased when cholera toxin was added either as a mixture or a conjugate to oral doses of Sendai virus (data not shown).

Table 1. Gut antibody titers (ELISA $\log_{10}$ vs. Sendai virus)[a]

Immunization	IgA[b]	IgG[b]
Virus alone	0.5 ± 0.3	0.5 ± 0.2
Mixture of virus plus cholera toxin	1.6 ± 0.8	0.9 ± 0.5
Conjugate of virus and cholera toxin	2.2 ± 0.6	1.1 ± 0.5

[a] Groups of 19 or 20 mice were immunized intragastrically on days 0, 14, 28, and 30 with 0.5 mg inactivated Sendai virus plus or minus 10 μg of cholera toxin either as a virus toxin mixture or covalent conjugate. Immunization was in a total volume of 0.5 ml of 0.2 N $NaHCO_3$. Mice were sacrificed on day 35 and antibody titers were determined in gastrointestinal contents (Nedrud et al. 1987).
[b] Geometric mean ± standard deviation.

When a comparable mass (13 μg) of cholera toxin B subunit was substituted for the holocholera toxin in either the virus-toxin mixture or conjugate, no enhancements of anti-viral antibody in gut washing or serum were observed (Liang et al. 1988). Thus, holocholera toxin but not cholera toxin B subunit was shown to be an effective mucosal adjuvant for oral Sendai virus. Potentiation of mucosal anti-horseradish peroxidase immune responses after enteric immunization with a covalent conjugate of B subunit and horseradish peroxidase observed by McKenzie and Halsey (1984) may have been due to a carrier effect of the B subunit; their ratio of B subunit to antigen was much higher than we used. In addition, theirs was a conjugate of B subunit to a soluble protein whereas we conjugated B subunit to a particulate antigen:inactivated virus particles.

Even with the addition of cholera toxin to intragastric doses of Sendai virus, however, little or no anti-viral antibody could be detected in either nasal washings or bronchoalveolar lavage fluids from orally immunized mice (Table 2 "oral"). Since we and other investigators have observed that local antigen deposition can enhance the accumulation and/or stimulate local proliferation of mucous membrane homing IgA lymphocytes (Pierce and Cray 1981, 1982; Weisz-Carrington et al. 1987), we boosted the orally immunized mice intranasally with marginally immunogenic doses of inactivated Sendai virus. Intranasal boosting resulted in the appearance of both IgA and IgG anti-viral antibodies in both nasal washings and bronchoalveolar lavage of orally immunized mice (Table 2 "oral and nasal"). In addition, both the upper and lower respiratory tracts of these orally primed, intranasally boosted mice were protected from infection by live virus (Fig. 1). Here 9 of 10 mice in the oral + nasal immunization group were totally protected from lung infection and 6 of 10 mice were totally protected from infection of the upper respiratory tract as well; the remaining four mice shed approximately 1000-fold less virus in nasal washings than controls. In contrast, mice which received only oral immunization were not protected from

Table 2. Virus and antibody titers in mice immunized orally, intranasally, or by combined route[a]

Immunization	Virus titer		Antibody titer					
			Nasal		Lung		Serum	
	Nasal	Lung	IgA	IgG	IgA	IgG	IgA	IgG
Control	5.2±1.5	4.3±1.3	---	---	---	---	---	0.4±0.6
Oral	5.6±0.6	4.8±1.1	---	---	---	0.2±0.2	1.9±1.0	2.5±0.6
Nasal	3.9±0.6	2.3±2.0	---	---	0.4±0.7	0.7±0.7	0.3±0.5	3.0±0.6
Oral & Nasal	1.9±0.6	1.7±1.0	1.5±0.6	0.6±0.4	3.2±0.6	3.0±0.4	3.2±0.7	4.4±0.7

[a]Mean antibody titers and virus titers (± standard deviation) for animals described in Fig. 1. Immunizations, and virus and antibody titers are per legends of Table 1 and Fig. 1. (--means no antibody detected). Reproduced from Nedrud et al. *J Immunol 139*:3484-3492, 1987. Copyright © 1987 by The American Association of Immunologists.

infection at either site and mice which received only the intranasal immunizations were variably protected: 6 of 10 were completely protected from lung infection and there was an average 10-fold reduction in virus titers in the upper respiratory tract. When the antibody titers in the various compartments were individually compared with virus titers for all the mice in this experiment, protection of the upper respiratory tract correlated best with nasal wash IgA, while lower respiratory tact protection correlated with bronchoalveolar lavage IgG. This comparison was done with computer software to compute stepwise linear regressions of antibody titer vs. protection for all 40 mice (Nedrud et al. 1987). Similar results have been obtained in two additional experiments, one employing a mixture of virus + cholera toxin and one employing virus-toxin conjugates for priming.

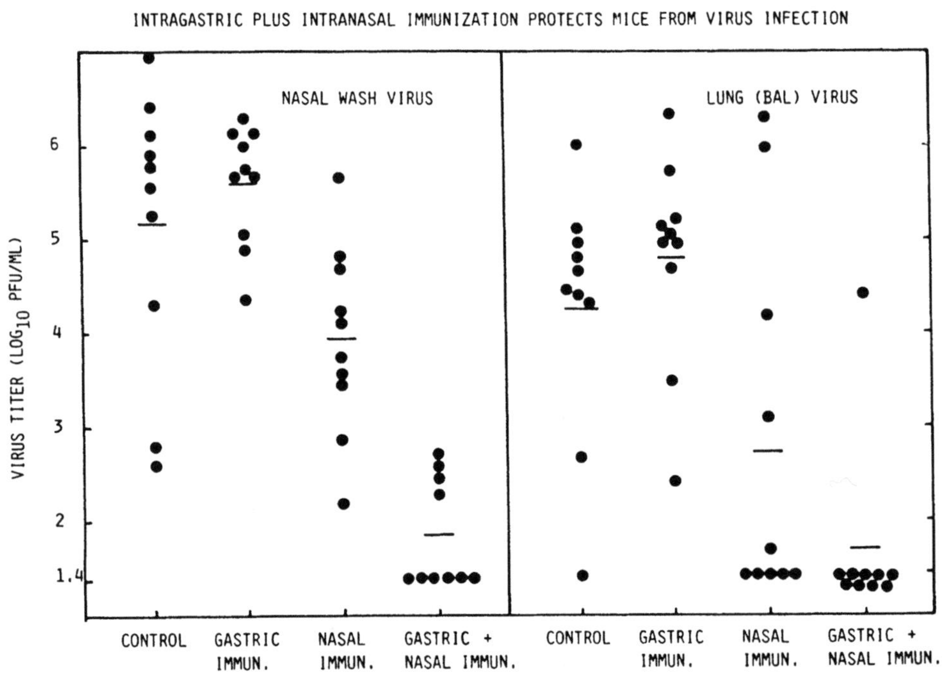

Fig. 1. Groups of 10 mice were intragastrically immunized according to the legend of Table 1. Some mice were intranasally immunized with 15μg of inactivated virus in saline on days 21 and 32, either alone or in addition to intragastric immunization. On day 35 the immunized mice plus an additional group of 10 control mice were intranasally challenged with 3.4 X 10^3 PFU (plaque forming units) of live Sendai virus while fully awake. Three days later the animals were killed and virus titers were determined for nasal washes and lung (bronchoalveolar lavage fluids). Each point represents a single animal and horizontal bars are the geometric mean virus titers for the group. The limit of detection in this assay was $10^{1.4}$ PFU/ml; all samples at or below this limit were set equal to 1.4 $\log_{10}$ PFU/ml for plotting and determination of means. Reproduced from Nedrud et al. J Immunol 139:3484-3492, 1987. Copyright© 1987 by The American Association of Immunologists.

These data are in agreement with other reports which have established a role for IgG (presumably derived from serum) in protection of the lower respiratory tract but which favor IgA (presumably locally synthesized) over IgG as significant in mediating resistance to infection of the upper respiratory tract. Experimental studies involving natural infection and/or vaccination invariably involve multiple types of antibody and cell-mediated immune response, however, so that it is difficult to ascribe protection from infection to any single mechanism. Passive transfer experiments have shown that under the proper conditions, serum, activated T cells, and monoclonal IgA as well as IgG antibodies can all mediate resistance to or recovery from respiratory viral infections (Lin and Askonas 1981; Mazanec et al. 1987; Örvell and Grandien 1982; Ramphal et al. 1979; Yap et al. 1978). Although numerous mechanisms are probably operative *in vivo*, results presented here and elsewhere (Mazanec et al. 1987) showed that intranasally applied anti-Sendai virus IgA monoclonal antibodies can passively protect mice from intranasal Sendai virus challenge. These results underscore the importance of peripheral IgA induction since IgA is protective and, at the same time, less phlogistic than IgG and many cell-mediated immune responses.

Since other investigators have achieved variable degrees of protection from infection without need for local boosting (Waldman et al. 1981; Clancy et al. 1985; Chen et al. 1987), it is possible that the local respiratory boosting requirement we observed is more of a technical problem than a theoretical requirement. We know, for instance, that Sendai virus is antigenically unstable at the pH of a mouse's stomach, so that even in the presence of an adjuvant like cholera toxin, the effective antigenic dose available to the intestinal lymphoid tissue may be quite low. This problem should be easier to circumvent in humans than in mice because enteric-coated capsules are readily available for human use.

Potentially more worrisome than the requirement for local boosting is the toxicity associated with the use of cholera toxin as an adjuvant. Although the dose used (10 µg) was quite low and little or no gross diarrhea was observed in mice used in the current studies, native cholera toxin would probably be unsuitable as an adjuvant for humans. Thus, we used higher concentrations of glutaraldehyde than employed earlier for the two-stage conjugate (Table 1) to prepare both free cholera toxoid and cholera toxoid covalently linked to Sendai virus via a single stage conjugation procedure. The toxicity in these latter preparations was reduced more than 1000-fold, but on a mass basis the Sendai virus-cholera toxoid conjugate was just as effective as Sendai virus mixed with native cholera toxin in enhancing gastrointestinal IgA anti-viral antibodies. The free cholera toxoid also retained mucosal adjuvant activity when mixed with oral Sendai virus but was less effective than the virus-toxoid conjugate (data not shown, Liang et al. submitted for publication, 1988).

SUMMARY

We have shown that in mice cholera toxin can be an effective adjuvant for gastrointestinal immune responses against a virus. The adjuvant properties can be increased and even dissociated from the toxic properties if virus and toxoid are covalently linked. Finally, oral immunization with these preparations of cholera toxin/toxoid and Sendai virus can be used to prime for respiratory immune responses to Sendai virus in which protection from infection correlates with IgA in the upper and with IgG in the lower respiratory tract.

REFERENCES

Chen K-S, Burlington DB, Quinnan Jr GV (1987) Active synthesis of hemagglutinin-specific immunoglobulin A by lung cells of mice that were immunized intragastrically with inactivated influenza virus vaccine. J Virol 61:2150-2154

Clancy R, Cripps A, Muree-Allen K, Yeung S, Engel M (1985) Oral immunization with killed *Hemophilus influenzae* for protection against acute bronchitis in chronic obstructive lung disease. Lancet ii:1395-1397

Elson CO, Ealding W (1984) Generalized systemic and mucosal immunity in mice after mucosal stimulation with cholera toxin. J Immunol 132:2736-2741

Liang X-P, Lamm ME, Nedrud JG (1988) Oral administration of cholera toxin-Sendai virus conjugate potentiates gut and respiratory immunity against Sendai virus. J Immunol 141:1495-1501

Lin Y-L, Askonas BA (1981) Biological properties of an influenza A virus-specific killer T cell clone. Inhibition of virus replication *in vivo* and induction of delayed-type hypersensitivity reactions. J Exp Med 154:225-234

Lycke M, Holmgren J (1986) Strong adjuvant properties of cholera toxin on gut mucosal immune responses to orally presented antigens. Immunology 59:301-308

Mazanec MB, Nedrud JG, Lamm ME (1987) IgA monoclonal antibodies protect against Sendai virus. J Virol 61:2624-2626

McKenzie SJ, Halsey JF (1984) Cholera toxin B subunit as a carrier protein to stimulate a mucosal immune response. J Immunol 133:1818-1824

Nedrud JG, Liang X-P, Hague N, Lamm ME (1987) Oral/nasal immunization protects mice from Sendai virus infection. J Immunol 139:3484-3492

Örvell C, Grandien M (1982) The effects of monoclonal antibodies on biologic activities of structural proteins of Sendai virus. J Immunol 129:2779-2787

Pierce NF, Cray Jr WC (1981) Cellular dissemination of priming for a mucosal immune response to cholera toxin in rats. J Immunol 127:2461-2464

Pierce NF, Cray Jr WC (1982) Determinants of the localization, magnitude and duration of a specific mucosal IgA plasma cell response in enterically immunized rats. J Immunol 128:1311-1315

Ramphal R, Cogliano RC, Shands Jr JW, Small Jr PA (1979) Serum antibody prevents lethal murine influenza pneumonitis but not tracheitis. Infect Immun 25:992-997

Rudzik R, Clancy RL, Perey DYE, Day RP, Bienenstock J (1975) Repopulation with IgA-containing cells of bronchial and intestinal lamina propria after transfer of homologous Peyer's patch and bronchial lymphocytes. J Immunol 114:1599-1604

Waldman RH, Lazzell VA, Bergmann C, Khakoo R, Jacknowitz AI, Howard SA, Rose C (1981) Oral immunization against influenza. Trans Am Clin Climatol Assoc 93:133-140

Weisz-Carrington P, Roux ME, McWilliams M, Phillips-Quagliata JM, Lamm ME (1979) Organ and isotype distribution of plasma cells producing specific antibody after oral immunization: Evidence for a generalized secretory immune system. J Immunol 123:1705-1708

Weisz-Carrington P, Grimes SR Jr, Lamm ME (1987) Gut-associated lymphoid tissue as source of an IgA immune response in respiratory tissues after oral immunization and intrabronchial challenge. Cell Immunol 106:132

Yap KL, Ada GL, McKenzie IFC (1978) Transfer of specific cytotoxic T lymphocytes protects mice inoculated with influenza virus. Nature 273:238-239

Immunity to Rotaviruses

R. Edelman[1], J. Flores[2], and A.Z. Kapikian[2]

[1]Clinical and Epidemiological Studies Branch, Microbiology and Infectious Diseases Program, and [2]Laboratory of Infectious Diseases, National Institute of Allergy and Infectious Diseases, NIH, Bethesda, MD 20892, USA

BACKGROUND: IMPORTANCE OF ROTAVIRUS

Diarrhea is the principal cause of mortality in infants and young children in the developing countries of Asia, Africa, and Latin America where the syndrome accounts annually for 3-5 billion cases and approximately 5-10 million deaths (Institute of Medicine 1985; Walsh and Warren 1979). Another estimate which reviewed data from several longitudinal studies in children revealed that 4.6 million diarrhea deaths and 744 million to 1 billion episodes of diarrhea occurred in children less than 5 years of age in developing countries, excluding China (Snyder and Merson 1982). In developed countries diarrheal deaths occur infrequently but morbidity from diarrheal illness is high (Institute of Medicine 1986; Rodriguez et al. 1987).

Rotaviruses are ubiquitous agents infecting virtually all children by 36 months of age. Since its discovery 14 years ago, rotavirus (RV) has been recognized as the single most common etiologic agent of infantile gastroenteritis leading to hospitalization in both developed and developing countries. For example, of 1537 children hospitalized with diarrheal illness in Washington, D.C. from 1974 to 1982, 34.5% shed RV in their stools (Brandt et al. 1983). Also 45% of children hospitalized with diarrhea in Yamagata City, Japan shed RV in their stools over a similar period of time (Konno et al. 1983). In Bangladesh, in a one-year study of patients coming to a treatment center with diarrheal illness, RV was the most frequently identified pathogen in children less than two years of age: 46% in this age group were RV-positive whereas 28% shed an enterotoxigenic *Escherichia coli*, the next most frequently detected agent (Black et al. 1980). By contrast, bacterial pathogens were identified more frequently than RV in children over two years of age (Black et al. 1980). On the other hand, rotavirus appears to induce severe dehydrating illness more frequently than other agents. For example, although RV was associated with <4% of all diarrheal episodes, RV was detected in 39% of illnesses that resulted in significant dehydration (Black et al. 1982). In total, 77% of episodes of diarrhea with life-threatening dehydration in children <2 years old were associated with RV or enterotoxigenic *E. coli* (Black et al. 1981). In the absence of fluid replacement therapy, Black et al. (1981) estimate that in Bangladesh RV mortality rate of 6.5/1000 individuals <2 years old could be expected. Because this virus is associated with a disproportionately high rate of life-threatening dehydrating episodes in developing countries and of hospitalizations in developed countries, aggressive efforts are being made to develop rotavirus vaccines for the world's infants and young children (De Zoysa and Feachem 1985). An efficacious RV vaccine administered to infants under 6 months of age in the developing countries may diminish the total number of cases of diarrhea by more than 50 million episodes and prevent 400,000 to 800,000 deaths per year (De Zoysa and Feachem 1985).

Current Topics in Microbiology and Immunology, Vol. 146

PATHOGENESIS OF HUMAN ROTAVIRUS IN ANIMALS

Since the human RV was first detected in 1973, information on the pathogenesis of RV infection has been gathered from studies in which the human virus was administered experimentally to domestic animals and monkeys. In a study by Mebus et al (1977), several calves were inoculated with human RV and sacrificed at intervals shortly after they developed diarrhea. Fluorescence-stainable viral antigens were found in villous epithelial cells in the lower small intestine, but the cells were morphologically normal. Within seven hours after diarrhea began, the RV antigens were no longer detected in the lower small intestine, but the villi were denuded and shortened. By 48 hours following onset of diarrhea the morphologic appearance had returned to normal. It was possible to produce illness only in calves inoculated by the second day of life. Calves inoculated 3 to 5 days after birth did not develop diarrhea (Wyatt et al. 1978).

Juvenile and 8 day-old Rhesus monkeys were susceptible to infection with human rotavirus (as determined by seroresponses with or without virus shedding), but diarrhea did not develop (Wyatt et al. 1976). Newborn Rhesus monkeys, however, did develop diarrhea after challenge. Four of the five monkeys that developed diarrhea had antibody in cord blood and RV particles were detected in the stool of each animal. Thus the diarrheal illness occurred in the presence of preexisting serum antibody. This finding is consistent with experimental evidence from piglets (Lecce et al. 1976), calves (Woode et al. 1975) and lambs (Snodgrass and Wells 1976), in which antibody at the mucosal surface was shown to be of major importance in resistance. The entire gastrointestinal tract of a newborn monkey was removed and studied 24 hours following administration of human RV (Wyatt et al. 1978). Viral antigens were identified by immunofluorescence in epithelial cells covering the villi of the upper through the lower small intestine, with little fluorescence in the lamina propria. Forty-five hours after RV infection, viral fluorescence had migrated away from epithelial cells and was found now principally in the lamina propria and mesenteric lymph nodes. These data suggest that human RV in Rhesus newborns is taken up and processed in the lymphoreticular system (Wyatt et al. 1978).

IMMUNITY TO ROTAVIRUS INFECTION

The role of the human immune system in protecting against RV infection is poorly understood. Because of marked differences in the local immune system from species to species, information obtained from animal studies is difficult to extrapolate to humans. Moreover, major differences exist between man and animal in the period over which they are susceptible to RV illness. In humans, the neonatal period is frequently asymptomatic, whereas in animals the neonatal period is the most critical time of vulnerability to RV illness. Nevertheless, in spite of these caveats, observations in animals taken together emphasize the fundamental concept that antibody in the intestinal lumen is necessary to afford protection against RV. Rotavirus illness in calves and lambs occurred in the presence of circulating RV antibody whereas RV antibody in the lumen of the small intestine was protective (Woode et al. 1975; Snodgrass and Wells 1976). Similarly, humans can develop illness from RV in the face of detectable RV serum antibodies (Bishop et al. 1983; Yolken et al. 1978a; Ward et al. 1986), and oral administration of RV antibody in gamma globulin to low birth weight newborns can delay and diminish the symptoms of rotavirus infection (Barnes et al. 1982). In a recent RV vaccine trial, children who had been previously infected developed a brisk, higher titered copro-IgA response with decreased and foreshortened shedding of the Rhesus RV vaccine; conversely a slow, primary copro-IgA response was associated with higher titered and prolonged RV shedding (Wright et al. 1987). Finally, RV vaccine can induce a coproantibody response without a detectible serum antibody response (Lonsonsky G, personal communication).

Although these observations cause us to focus attention on local immune factors as mediators of protection, in man the precise mechanisms of induction and role of mucosal immunity in control and prevention of illness is unclear (Riepenhoff-Talty et al. 1981; Kapikian et al. 1983; Hjelt et al. 1986). Until the mechanisms by which RV induces an immune response in the intestine are better understood and the methods of measuring the response are better developed (Yamaguchi et al. 1985) we must depend on measurements of serum antibody. It is likely that changes in serum antibody titers, in particular RV-neutralizing antibody, may only reflect events at the mucosal level and that these events are the actual effectors of immunity. However, it is also possible that transfer of neutralizing antibody to intestinal secretions from plasma could occur and enhance viral neutralization and mucosal immunity. In fact, circulating antibody may be sustained even after local antibody has disappeared, although immunologic memory in the bowel may persist (Wright et al. 1987).

Until we can reliably measure mucosal immunity, longitudinal observations of the serum antibody response to natural RV infections are urgently needed. In the one such study published, Chiba et al. (1986) reported that a serum-neutralizing antibody titer of 1:128 or greater to type 3 rotavirus was associated with protection of Japanese children against subsequent serotype 3 infection and illness. Heterologous antibody ($\geq$ 1:128) to serotypes 1, 2, and 4 was not associated with protection against serotype 3. In other epidemiological studies subclinical and symptomatic infections were observed in adult contacts of infants with RV gastroenteritis, but exposed children were more likely to acquire RV infection and become symptomatic than adults in the same family (Greenwood et al. 1983; Rodriguez et al. 1987). It is unknown whether symptomatic reinfections with RV are the result of shortened duration of the immune response to the initial infection, or of susceptibility to other rotavirus serotypes different from the one causing the initial illness. In another study, a group of children infected during the neonatal period had similar rates of RV reinfection after three years of follow-up compared to uninfected neonates; however, less symptomatic illness was observed in the neonatally infected children when reinfection occurred later in life (Bishop et al. 1983). The sum of the epidemiological observations suggest that protection against RV disease may be conferred by the administration of vaccines early in life. The study by Chiba et al. (1986) hints that such vaccines need to be serotype-specific.

ROTAVIRUS STRUCTURE AND SEROTYPES

A substantial amount of information is available on the biochemical structure of the virus and its genome (Fig. 1-taken from Kapikian et al. 1986). Such knowledge is basic to the understanding of immunity to rotaviruses. The rotaviruses are double shelled 70 nm particles formed by at least five proteins. Two such proteins, VP-7 (a 38 kDal glycoprotein) and VP-3 (an 88 kDal protein), form the outer surface of the virus and are capable of eliciting neutralizing antibodies against the virus. Another major antigenic component not related to virus neutralization is VP-6 (a 42 kDal protein) located in the inner shell of the virion. This protein comprises the common RV group antigen and the epitopes that define the two RV subgroups (Flores et al. 1986b).

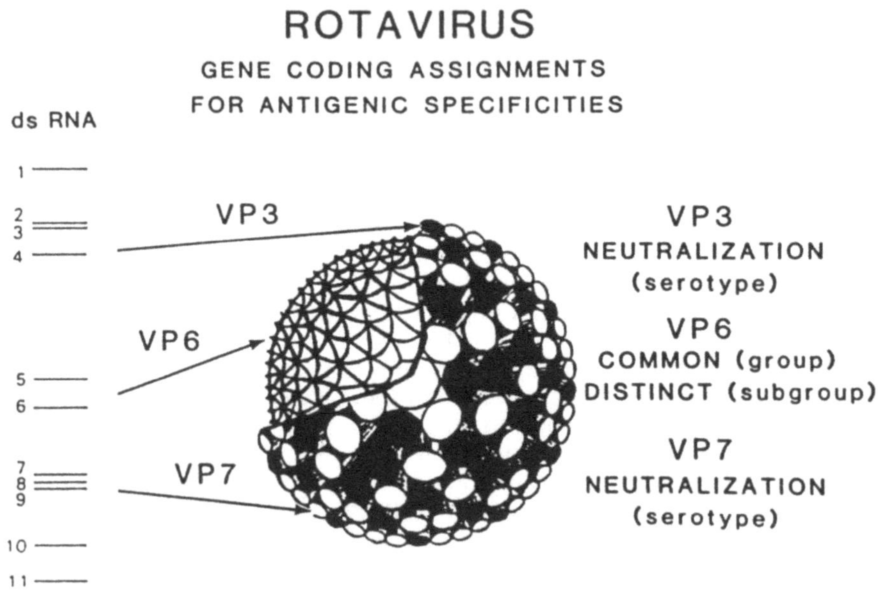

Fig. 1. Rotavirus - gene coding assignments for antigenic specificities.

The RV genome consists of 11 segments of double stranded RNA. Each segment acts as an individual gene directing the synthesis of one protein. Virus neutralization tests have clearly identified at least six human serotypes (Wyatt et al. 1983a; Hoshino et al. 1984; Albert et al. 1987; Clark et al. 1987a), but serotypes 1-4 seem to be the most important epidemiologically. VP-7, encoded by genomic RNA segment 8 or 9, represents the major neutralizing, serotype-specific epitope or epitopes (Kalica et al. 1981; Midthun et al. 1986). The VP-7 of each RV serotype differs sufficiently in its amino acid sequence to account for the antigenic diversity found among rotavirus serotypes (Glass et al. 1985; Green et al. 1987). VP-3, encoded by genomic RNA segment 4, is currently thought to modulate the difference in virulence between RV strains (Flores et al. 1986a; Gorziglia et al. 1986) and to play a role in virus neutralization (Hoshino et al. 1985a; Offit and Blavat 1986).

Unfortunately, only a limited number of research laboratories are able to serotype rotaviruses isolated from human fecal samples and to measure serotype-specific RV antibody in human and intestinal washings. The current methods are laborious and time-consuming and the reagents are in short supply (Gerna et al. 1985). The recent development, both of monoclonal antibodies to react specifically with each of the four important human serotypes of VP-7 (Greenberg et al. 1983; Taniguchi et al. 1985; Shaw et al. 1985; Coulson et al. 1987; Birch et al. 1988) and of RNA-RNA hybridization to differentiate serotypes by identification of their genetic differences (Flores et al. 1986a; and unpublished results) should help overcome the technical barriers to widespread serotyping. In addition, a VP-7 serotype-specific serologic assay has been developed recently (Shaw et al. 1987).

Because few RV strains have been cultured and serotyped, estimates of the relative prevalence of any serotype are quite imprecise. To date scanty epidemiologic information suggests that serotype 1 is consistently the most prevalent with the distribution of the other 3 serotypes depending upon study location and temporal factors (Gerna et al. 1985).

Although the RV genome is relatively resistant to point mutations (Flores et al. 1986b), gene reassortment among RV appears to occur under natural conditions (Flores et al. 1985a; Flores et al. 1986b). Reassortant events may account for the high degree of genetic variability observed in RV (Flores et al. 1982; Rodger et al. 1981), and may allow the generation of RV strains with new serotype specificities. The extent and clinical importance of such reassortment, however, is currently unknown.

HOMOTYPIC IMMUNITY

Although both homotypic and heterotypic immunity may be elicited by infection, it is not known how long infection with one RV serotype can protect against RV reinfection and/or illness by its own serotype, and how broad and lasting is the cross-protection afforded by one serotype against other serotypes. In terms of serotype-specific immunity, studies in gnotobiotic calves showed that cross-protection occurred only between rotaviruses of the same serotype and that even a minor serotype difference eliminated their ability to cross-protect (Woode et al. 1983). The evidence for homotypic immunity in man is currently limited to one epidemiological study (Chiba et al. 1986) and one vaccine trial (Flores et al. 1987). The study by Chiba et al. (1986) showed that homotypic serum antibody titers of 1:128 or greater, as measured by plaque neutralization assays, were generally associated with protection of children exposed to an outbreak of serotype 3 rotavirus. Pre-existing heterotypic antibody titers of up to 1: 512 against serotypes 1, 2 and 4, or homotypic titers less than 1:128, were not generally associated with protection. In the recent vaccine trial performed in Caracas, Venezuela, the Rhesus rotavirus vaccine (serotype 3), provided an overall protective efficacy of 68% in 1-10 month old infants during one year of continuous follow-up (Flores et al. 1987). Among the 1-5 months old infants, vaccine efficacy was 93%, and it was 100% efficacious against the most severe diarrheal episodes. Analysis of the serotypes isolated during the trial revealed that serotype 3 strains were the most prevalent, indicating that the protection observed was mostly homotypic. The serotype 3 Rhesus rotavirus vaccine was also tested in infants living in Rochester, New York and on the Navajo Indian Reservation in Arizona (Table 1). The vaccine failed to protect infants in these two trials, both of which were principally associated with serotype 1 infection. Although these trial failures suggest a lack of heterotypic immunity, the same vaccine was at least partially efficacious in Finland and Sweden against principally serotype 1 infections (Table 1) thereby suggesting some role for heterotypic immunity in man. Such immunity is further discussed below.

HETEROTYPIC IMMUNITY

Despite the absence of sound epidemiologic data on the mechanism of human resistance to RV serotypes, three candidate vaccines derived from animal strains, RIT 4237, RRV-1 and WC3 have progressed to field trials (Table 1). They provide insight into the role of heterospecific immunity in man. By way of background, efforts to develop attenuated strains of human rotaviruses for vaccines have been hampered by the poor cultivability of human strains of RV and the difficulty in determining the pathogenic potential of different strains of human virus (Kapikian et al. 1986b). Also it was assumed that a degree of heterotypic immunity would develop, as was seen in animal models (Wyatt et al. 1979; Wyatt 1983; Zissis et al. 1983) and in volunteers (Kapikian et al. 1983). Thus heterologous, antigenically cross-reactive RV strains derived from animal sources have been developed for vaccine use.

Table 1. Status of monovalent rotavirus efficacy vaccine trials completed or underway as of March 1, 1988

Candidate vaccine	Principal investigator	Location	Efficacy against rotavirus	Reference
RIT 4237	Vesikari	Finland	+	1984b
(bovine,	Vesikari	Finland	+	1985
serotype 6)	Vesikari	Finland	+	1987a
	De Mol	Rwanda	0	1986
	Hanlon	Gambia	0	1987
	Senturia	London, England	0	1987
	Santosham	Arizona, USA	0	P
	Lanata	Peru	+	P
	Yolken	Baltimore, USA	RP	P
RIT 4256 (bovine, serotype 6)	Vesikari	Finland	SA	1987b
WC3 (bovine,	Clark	Philadelphia, USA	++	1987a,b
serotype 6)	Alpert	Israel	RP	P
RRV-1	Flores	Venezuela	++	1987
(Rhesus	Vesikari	Finland	+	P
monkey,	Gothefors	Sweden	+	P
serotype 3)	Dolin	Rochester, USA	0	P
	Santosham	Arizona, USA	0	P

0, none; + against severe diarrhea only; ++ against all diarrhea; RP, results pending; SA, safety and antigenicity trials only; P, personal communication.

These trials have shown that heterotypic vaccines derived from animal RV strains confer inconsistent protection (Table 1). In results to date the vaccines either: (1) protect against mild and severe RV gastroenteritis and decrease the incidence of RV illness overall; (2) reduce the incidence of severe gastroententis without reducing the incidence of RV illness overall; or (3) do not prevent any RV illness. If there is a role for heterotypic antibody it may result from a RV epitope or epitopes shared by different RV types located on VP3 (Taniguchi et al. 1985; Clark et al. 1987a). Theoretically human infants may be primed by infection with such an epitope(s) and have antibody to it boosted after repeated infection with other RV serotypes. This boost may result in heterospecific antibody which has some neutralizing capacity (Kapikian et al. 1986a) and might provide some ancillary protection against all serotypes that carry the common VP3 antigen on their surface (Offit et al. 1986c). The existence of similar cross reactive epitopes on VP7 has not yet been demonstrated.

MULTISEROTYPE VACCINES

Evidence to date suggests that if vaccines are to work in all clinical settings, multivalent preparations containing several or all of the prevalent RV serotypes will be required.

In-other-words, serotype-specific immunity may be more efficacious than heterospecific immunity. In addition to the human studies cited above, most animal studies show that neutralizing antibody against rotavirus is highly serotype-specific and that heterotypic antibodies do not provide significant degrees of cross-neutralization *in vitro* (Hoshino et al. 1984; Greenberg et al. 1983) or protection *in vivo* (Woode et al. 1983; Offitt and Clark 1986b; Losonsky et al. 1986; Hoshino et al. 1987).

The first attempts to develop and test multivalent vaccines are underway (Kapikian et al. 1986b). In a system designed to produce reassortant rotaviruses with the serotype specificities of human rotaviruses, fastidious human rotaviruses have been co-cultivated with less fastidious animal rotavirus strains in tissue culture. In these experiments, 10 segmented genes of the noncultivatable human rotavirus are replaced by the 10 animal genes (bovine or Rhesus monkeys). The single remaining human rotavirus gene (gene 8 or 9) coding for VP7, the major neutralizing protein on the virus surface, is preserved. Single gene substitution in RV reassortants have been prepared for human serotypes 1, 2, 3 and 4 with the Rhesus rotavirus strain or the UK bovine strain (Midthun et al. 1985, 1986). In addition, the WC3 bovine strain has been reassorted with the human RV strains of types 1 and 3 (Clark et al. 1987b). Other donor strains, such as naturally attenuated "nursery" strains (Bishop et al. 1983; Flores et al. 1986a) and cold adapted human strains (Matsuno et al. 1987) are being considered as reassortant donors. Theoretically, such gene substitution reassortants should induce immunity to viruses belonging to the VP7 serotype of their human RV parent, while retaining the attenuation of their animal or human RV parent. The attenuated parents should allow viral growth to high titer *in vitro*, thus making the reassortants suitable for economic vaccine production. If in the future, additional, clinically important rotavirus serotypes emerge, they too could be attenuated by transferring the genes of the attenuated donor strain to reassortants bearing the VP7 gene of the newly recognized serotype. If the single gene substitution reassortants prove less than satisfactory, it may be necessary to include gene 4 which codes for VP3, the less dominant and more widely shared neutralizing protein on the virus surface (Hoshino et al. 1985), provided it is itself attenuated (Flores et al. 1986a). Safety, antigenicity and efficacy trials of VP7 reassortant rotaviruses tested separately and as combined vaccines are underway (Table 2).

Also available are seemingly naturally attenuated strains isolated from asymptomatic neonates (Bishop et al. 1983; Perez-Schael et al. 1984; Flores et al. 1986ba; Gorziglia et al. 1986). Such strains, represented by each of the four important human serotypes 1-4 (Hoshino et al. 1985b) could be tested individually and then together as a multivalent vaccine preparation. The newly developed, cold-attenuated human strains offer similar possibilities as vaccine candidates (Matsuno et al. 1987).

OTHER VACCINES ISSUES

The reader should appreciate from the foregoing discussion that RV vaccinology is complex. In addition, a number of other important clinical and epidemiological issues, both real and theoretical, must be resolved before RV vaccines are successfully developed and deployed. Some of these issues are outlined below.

Table 2. Status of reassortant Rotavirus vaccine trials completed or underway as of March 1, 1988

Candidate vaccine	Principal investigator	Location	Efficacy against rotavirus	Reference
RRV-1 x D (Rhesus-human, serotype 1)	Halsey/ Clements	Baltimore, USA	SA	P
	Wright	Nashville, USA	SA	P
	Vesikari	Finland	RP	P
	Dolin	Rochester, USA	RP	P
	Santosham	Arizona, USA	RP	P
	Rennels	Baltimore, USA	RP	P
	Glass	Atlanta, USA	SA	P
	Lanata	Peru	RP	P
	Perez-Schael	Venezuela	SA	P
RRV-1 x DS1 (Rhesus-human, serotype 2)	Anderson	Huntington, USA	SA	P
	Vesikari	Finland	RP	P
	Lanata	Peru	RP	P
	Halsey	Baltimore, USA	SA	P
	Perez-Schael	Venezuela	SA	P
RRV-1 x ST3 (Rhesus-human, serotype 4)	Halsey	Baltimore, USA	SA	P
RRV-1 x D RRV-1 X DS1 RRV-1 RRV-1 x ST3 (rhesus-human, Serotype 1 + 2 + 3 + 4)	Flores/ Perez-Schael	Venezuela	RP	P
WC3:2-5 (bovine-human; serotype 3)	Clark	Pennsylvania, USA	SA	1987b
WC3: WI79-9 (bovine-human, serotype 1)	Clark	Pennsylvania, USA	SA	1987b

RP, results pending; SA, safety and antigencity trials only; P, personal communication.

Grading the Severity of Illness

In order to establish vaccine efficacy it will be essential to develop a clinical grading system that allows objective comparison of gastroenteritis occurring in vaccine and placebo treated children. Moreover, we still lack details of the myriad clinical presentations of RV infections occurring in different geographic and socioeconomic settings. Our knowledge of RV disease rests mostly on studies of children taken to hospitals and clinics. Most clinical reports are therefore biased toward the more severely ill child (Rodriguez et al. 1977; Flewett 1981). We need detailed descriptions of community-based RV disease to elucidate its full clinical spectrum accompanied by serologic and virologic studies to detect inapparent infections (Rodriguez et al. 1987).

Oral Polio Vaccines

For logistic reasons it would be essential in many countries to administer oral RV vaccines concomitantly with the oral polio vaccine. Anecdotal and controlled data (Vodopija et al. 1986) suggest that enteroviruses interfere with RV vaccine in a manner similar to the interference among polio strains in the trivalent oral polio vaccine (Patriarca et al. 1988). Of even greater importance to public immunization practice would be to determine if RV vaccine inhibits polio vaccine. If confirmed and uncorrected, such interference would render simultaneous administration of oral polio and RV vaccines impractical (Wright et al. 1987).

Secondary Transmission of Vaccine Virus

Vaccine virus shed in the feces would be an advantage if others would be immunized by fecal-oral transmission, and if the vaccine strain remained stable genetically. Because frequent viral shedding has been documented in the stools of Rhesus RV vaccines, it will be important to continue to measure the titers and genetic characteristics of the virus shed and rates if any, of secondary transmission.

Vaccine Buffers and Human Milk

Because rotaviruses are susceptible to acid (Vesikari et al. 1984a) the effect of different gastric acid buffers on the potency of RV vaccine needs to be determined. Breast-feeding might enhance vaccine potency due to its gastric buffering action (Vesikari et al. 1984a), or it might inhibit RV vaccine because of anti-rotavirus antibody (Yolken et al. 1978b; Berger et al. 1984) and nonspecific RV inhibitors (McLean and Holmes 1981). Any impediment to vaccine "take" by breast-milk will complicate utilization of the vaccine in nursing infants and reduce its acceptance.

Vaccine Dosage

A one-dose vaccine is ideal logistically, but a single dose may fail to infect consistently. Some of the many possible causes of oral RV vaccine failure is outlined in Table 3. This failure, together with the transient nature of intestinal fluid antibody which wanes rapidly compared with serum antibody (Kapikian et al. 1983), implies the need for either repeated vaccine doses or frequent natural exposure to wild virus to maintain mucosal immunity. Multiple dose regimens that can be easily accommodated in existing immunization program schedules should be systematically investigated.

Table 3. Some causes of oral rotavirus vaccine failure

Failure to swallow the vaccine or emesis.
Gastric acid.
Presence of maternal transplacental or milk antibody in the infant.
Antibody present from previous RV infection.
Interference by replicating enteroviruses in the bowel.
Age-related changes in cellular viral receptors.
Rapid bowel transit time.

Concurrent Infections by Rotavirus and Other Enteropathogens

In some overseas settings, as many as 40% of patients with diarrhea shed RV and at least one other enteropathogen concurrently (Mata et al. 1983). Such mixed infections should be sought in vaccine trials in order to: 1) establish the likelihood that an RV vaccine would provide meaningful protection in the face of intense transmission of other enteric pathogens; 2) establish vaccine efficacy for RV infections; and 3) determine if rotaviruses enhance the virulence of bacterial infections in man as they do in animals (Lecce et al. 1982; Moreau 1986). If so, an effective RV vaccine might reduce diarrhea morbidity more than would be predicted by its effect on RV alone.

SUMMARY AND CONCLUSION

A potent, multivalent, serotype-specific RV vaccine and improved tests for measuring vaccine potency would help eliminate the necessity to pretest for vaccine efficacy in every country selected for its deployment. Until then, the need will continue for vaccine trials in various countries because the pathogenesis and epidemiology of RV and RV serotypes differ between and within countries. Although RV vaccinology is complex, it has forged ahead of our knowledge of RV immunopathogenesis and epidemiology.

REFERENCES

Albert MJ, Unicomb LE, Bishop RF (1987) Cultivation and characterization of human rotaviruses with "super short" RNA patterns. J Clin Microbiol 25:183-185

Barnes GL, Doyle LW, Hewson PH, Knocles AML, McLellan JA, Kitden WH, Bishop RF (1982) A randomized trial of oral gammaglobulin in low birth weight infants infected with rotavirus. Lancet i:1371-1373

Berger R, Hadziselimovic F, Just M, Reigel F (1984) Influence of breast milk on nosocomial rotavirus infection in infants. Infection 12:171-174

Birch CJ, Heath RL, Gust ID (1988) Use of serotype-specific monoclonal antibodies to study the epidemiology of rotavirus infection. J Med Virol 24:45-53

Bishop RF, Barnes GL, Cipriani E, Lund JS (1983) Clinical immunity after neonatal rotavirus infection: A prospective longitudinal study in young children. N Engl J Med 309:72-76

Black RE, Merson MH, Rahman ASMM, Yunus M, Alim ARMA, Huq I, Yolken RH, Curlin GT (1980) A two-year study of bacterial, viral, and parasitic agents associated with diarrhea in rural Bangladesh. J Infect Dis 5:660-664

Black RE, Merson MH, Huq I, Alim ARMA, Yunus M (1981) Incidence and severity of rotavirus and *Escherichia coli* diarrhea in rural Bangladesh: Implications for vaccine development. Lancet i:141-143

Black RE, Brown KH, Becker S, Alim ARMA (1982) Longitudinal studies of infectious diseases and physical growth of children in rural Bangladesh. II. Incidence of diarrhea and association with known pathogens. Am J Epidemiol 115:315-324

Brandt CD, Kim HW, Rodriguez WJ, Arrobio JO, Jeffries BC, Stallings EP, Lewis C, Miles AJ, Chanock RM, Kapikian AZ, Parrott RH (1983) Pediatric viral gastroenteritis during eight years of study. J Clin Microbiol 18:71-78

Chiba S, Yokoyama T, Nakata S, Morita K, Urasawa T, Taniguchi K, Urasawa S, Nakao T (1986) Protective effect of naturally acquired homotypic and heterotypic rotavirus antibodies. Lancet ii:417-421

Clark HF, Furukawa T, Bell LM, Offit PA, Perrella PA, Plotkin SA (1986) Immune response of infants and children to low passage bovine rotavirus (strain WC3). Am J Dis Child 140:350-356

Clark HF, Hoshino Y, Bell LM, Groff J, Hess G, Brachman P, Offit PA (1987a) Rotavirus isolate W161 representing a presumptive new human serotype. J Clin Microbiol 25:1757-1762

Clark HF, Borian FE, Modesto K, Rodden J, Bell LM, Plotkin SA (1987b) Recent progress in evaluating rotavirus vaccines for human infants in Program of the US-Japan virology panel meeting, December 1-4, Monterey, California (abstract)

Coulson BS, Unicomb LE, Pitson GA, Bishop RF (1987) Simple and specific enzyme immunoassay using monoclonal antibodies for serotyping human rotavirususes. J Clin Microbiol 25:509-515

De Mol P, Zissis G, Butzler JP, Mutwewingabo A, Andre FE (1986) Failure of live, attenuated oral rotavirus vaccine. Lancet ii:108

De Zoysa I, Feachem RG (1985) Interventions for the control of diarrhoeal diseases among young children: Rotavirus and cholera immunization. Bull WHO 63:569-583

Flewett TH (1981) Clinical features of rotavirusus infection. In: Tyrrell DAJ, Kapikian AZ (eds) Virus Infections of the Gastrointestinal Tract. New York, Marcel Dekker, p 125

Flores J, Perez I, White L, Perez MA, Kalica AR, Marquina R, Wyatt RG, Kapikian AZ, Chanock RM (1982) Genetic relatedness among human rotaviruses as determined by RNA hybridization. Infect Immunity 37:648-655

Flores J, Perez-Schael I, Boeggeman E, White L, Perez M, Purcell R, Hoshino Y, Midthun K, Chanock RM, Kapikian AZ (1985a) Genetic relatedness among human rotaviruses. J Med Virol 17:135-143

Flores J, Midthun K, Hoshino Y, Green K, Gorziglia M, Kapikian AZ, Chanock RM (1986a) Conservation of the fourth gene among rotaviruses recovered from asymptomatic newborn infants and its possible role in attenuation. J Virol 60:972-979

Flores J, Nakagomi O, Nakagomi T, Glass R, Gorziglia M, Askaa J, Hoshino Y, Perez-Schael I, Kapikian AZ (1986b) The role of rotaviruses in pediatric diarrhea. Pediatr Infect Dis 5:S53-S62

Flores J, Perez-Schael I, Gonzalez M, Garcia D, Perez M, Daoud N, Cunto W, Chanock RM, Kapikian AZ (1987) Protection against severe rotavirus diarrhoea by Rhesus rotavirus vaccine in Venezuelan infants. Lancet i:882-884

Gerna G, Assarani N, Saraini A, Battaglia M (1985) Characterization of serotyes of human rotavirus strains by solid-phase immune electron microscopy. J Infect Dis 152:1143-1151

Glass R, Keith J, Nakagomi O, Nakagomi T, Askaa J, Kapikian AZ, Chanock RM, Flores J (1985) Nucleotide sequence of the structural glycoprotein VP-7 gene of Nebraska calf diarrhea virus rotavirus: Comparison with homologous genes from four strains of human and animal rotaviruses. Virology 141:292-298

Gorziglia M, Hoshino Y, Buckler-White A, Blumentals I, Glass R, Flores J, Kapikian AZ, Chanock RM (1986) Conservation of amino acid sequence of VP8 and cleavage region of 84-KDa outer capside protein among rotaviruses recovered from asymptomatic neonatal infection. Proc Natl Acad Sci USA 83:7039-7043

Green KY, Midthun K, Gorziglia M, Hoshino Y, Kapikian AZ, Chanock RM, Flores J (1987) Comparison of the amino acid sequences of the major neutralization protein of four human rotavirus serotypes. Virology 161:153-159

Greenberg HB, Valdesuso J, van Wyke K, Midthun K, Walsh M, McAuliffe V, Wyatt RG, Kalica AR, Flores J, Hoshino Y (1983) Production and preliminary characterization of monoclonal antibodies at two surface proteins of Rhesus rotavirus. J Virol 47:267-275

Grimwood K, Abbott GD, Fergusson DM, Jennings LC, Allan JM (1983) Spread of rotavirus within families: A community-based study. Br Med J 287:575-577

Hanlon P, Hanlon L, Marsh V, Byass P, Shenton F, Hassan-king M, Jobe O, Sillah H, Hayes R, M'Boge BH, Whittle HC, Greenwood BM (1987) Trial of an attenuated bovine rotavirus vaccine (RIT 4237) in Gambian infants. Lancet i:1342-1345

Hjelt K, Grauballe PC, Andersen L, Schiotz PO, Howitz P, Krasilnikoff PA (1986) Antibody response in serum and intestine in children up to six months after a naturally acquired rotavirus gastroenteritis. J Pediatr Gastroenterol Nutr 5:74-80

Hoshino Y, Wyatt RG, Greenberg HB, Flores J, Kapikian AZ (1984) Serotypic similarity and diversity of rotaviruses of mammalin and avian origin as studied by plaque reduction neutralization. J Infect Dis 149:694-702

Hoshino Y, Sereno MM, Midthun K, Flores J, Kapikian AZ, Chanock RM (1958a) Independent segregation of two antigenic specificities (VP3 and VP7) involved in neutralization of rotavirus infectivity. Proc Natl Acad Sci USA 82:8701-8704

Hoshino Y, Wyatt RG, Flores J, Midthun K, Kapikian AZ (1985b) Serotype characterization of rotaviruses derived from asymptomatic human neonatal infection. J Clin Microbiol 21:425-430

Hoshino Y, Torres A, Aves TM, Flores J, Lin J, Sereno MM, Kapikian AZ (1987) Neutralization antibody response of gnotobiotic piglets after fetal infection and neonatal challenge with human rotavirus strains. J Infect Dis 156:1038-1040

Institute of Medicine (1985) New Vaccine Development, Establishing Priorities. Vol I. Washington, DC, National Academy Press

Institute of Medicine (1986) New Vaccine Development. Establishing Priorities. Vol II. Washington, DC, National Academy Press

Kalica AR, Greenberg HB, Wyatt RG, Flores J, Sereno MM, Kapikian AZ, Chanock RM (1981) Genes of human (strain WA) and bovine (strain UK) rotaviruses that code for neutralization and subgroup antigens. Virology 112:385-390

Kapikian AZ, Wyatt RG, Levine MM, Yolken RH, Vankirk DH, Dolin R, Greenberg HB, Chanock RM (1983) Oral administration of human rotavirus to volunteers: Induction of illness and correlates of resistance. J Infect Dis 147:95-106

Kapikian AZ, Hoshino Y, Flores Y, Midthun K, Glass RI, Nakagomi O, Nakagomi T, Chanock RM, Potash L, Levine MM, Dolin R, Wright PF, Belshe RE, Anderson EL, Vesikari T, Gothefors L, Wadell G, Perez-Schael I (1986a) Alternative approaches to the development of a rotavirus vaccine. In: Holmgren J, Lindberg A, Mollby R (eds) Development of Vaccines and Drugs Against Diarrhea, 11th Nobel conference, Stockholm 1985. Lund, Sweden, Studentlitteratur, p 192

Rodriguez WJ, Kim HW, Arrobio O, Brandt CD, Chanock RM, Kapikian AZ, Wyatt R, Parrott RH (1977) Clinical features of acute gastroenteritis associated with human reo-virus-like agent in infants and young children. J Pediatr 91:188-193

Rodriguez WJ, Kim HW, Brandt CD, Schwartz RH, Gardner MK, Jeffries B, Parrott RH, Kaslow RA, Smith JI, Kapikian AZ (1987) Longitudinal study of rotavirus infection and gastroenteritis in families served by a pediatric medical practice: Clinical and epidemiological observations. Pediatr Infect Dis J 6:170-176

Senturia YD, Peckham CS, Cordery M, Chrystie IA, Banatvala JE, Andre FE (1987) Live attenuated oral rotavirus vaccine [letter]. Lancet ii:1091-1092

Shaw RD, Stoner-Ma DL, Estes MK, Greenberg HB (1985) Specific enzyme-linked immunoassay for rotavirus serotypes 1 and 3. J Clin Microbiol 22:286-291

Shaw RD, Fong KJ, Losonsky GA, Levine MM, Maldonado Y, Yolken R, Flores J, Kapikian AZ, Vo PT, Greenberg HB (1987) Epitope-specific immune responses to rotavirus vaccination. Gastroenterology 93:941-950

Snodgrass DR, Wells PW (1976) Rotavirus infection in lambs: Studies on passive protection. Arch Virol 52:201-205

Snyder JD, Merson MH (1982) The magnitude of the global problem of acute diarrhoeal disease: A review of active surveillance data. Bull WHO 60:605-613

Taniguchi K, Urasawa S, Urasawa T (1985) Preparation and characterization of neutralizing monoclonal antibodies with different reactivity patterns to human rotaviruses. J Gen Virol 66:1045-1053

Vesikari T, Isolauri E, D'Hondt E, Delem A, Andre FE (1984a) Increased take rate of oral rotavirus vaccine in infants after milk feeding. Lancet ii:700

Vesikari T, Isolauri E, D'Hondt E, Delem A, Andre F, Zissis G (1984b) Protection of infants against rotavirus diarrhea by RIT 4237 attenuated bovine rotavirus vaccine strain. Lancet i:977-981

Vesikari T, Isolauri, E, Delem A, D'Hondt E, Andre FE, Beards GM, Flewett TH (1985) Clinical efficacy of the RIT 4237 live attenuated bovine rotavirus vaccine in infants vaccinated before a rotavirus epidemic. J Pediatr 107:189-194

Vesikari T, Ruuska T, Delem A, Andre FE (1987a) Neonatal rotavirus vaccination with RIT 4237 bovine rotavirus vaccine: A preliminary report. Pediatr Infect Dis J 6:164-169

Vesikari T, Rautanen T, Isolauri E, Delem A, Andre FE (1987b) Immunogenuity and safety of a low passage level bovine rotavirus candidate vaccine RIT 4256 in human adults and young infants. Vaccine 5:105-108

Vodopija I, Baklaic Z, Vlatkovic R, Bogaerts H, Delem A, Andre FE (1986) Combined vaccination with live oral polio vaccine and the bovine rotavirus RIT 4237 strain. Vaccine 4:233-236

Walsh JA, Warren KS (1979) Selective primary health care: An interim strategy for disease control in developing countries. N Eng J Med 301:967-974

Ward RL, Bernstein DI, Young EC, Sherwood DI, Knowlton DR, Schiff GM (1986) Human rotavirus studies in volunteers: Determination of infections dose and serological response to infection. J Infect Dis 154:871-879

Woode GN, Jones J, Bridger JC (1975) Levels of colostral antibodies against neonatal calf diarrhea virus. Vet Rec 97:148-149

Woode GN, Kekelso NE, Simpson TF, Gaul SK, Evans LE, Babiuk L (1983) Antigenic relationships among some bovine rotaviruses: Serum neutralization and cross-protection in gnotobiotic calves. J Clin Microbiol 18:358-364

Wright PF, Tajima T, Thompson J, Kokubun K, Kapikian AZ, Karzon DT (1987) Candidate rotavirus vaccine (Rhesus rotavirus strain) in children: An evaluation. Pediatrics 80:473-480

Wyatt RG, Sly DL, London WT, Papmer AE, Kalica AR, van Kirk DH, Chanock RM, Kapikian AZ (1976) Induction of diarrhea in colostrum-deprived newborn Rhesus monkeys with human reovirus-like agent of infantile gastroenteritis. Arch Virol 50:17-27

Wyatt RG, Kalica AR, Mebus CA, Kim HW, London WT, Chanock RM, Kapikian AZ (1978) Reovirus like agents (Rotaviruses) associated with diarrheal illness in animals and man. In: Pollard M (ed) Perspectives in Virology, Vol 10, Raven Press, New York, p 121

Wyatt RG, Mebus CA, Yolken RH, Kalica AR, James Jr HD, Kapikian AZ, Chanock RM (1979) Rotavirus immunity in gnotobiotic claves: Heterologous resistance to human virus induced by bovine virus. Science 203:548-550

Wyatt RG, James Jr HD, Pittman AL, Hoshino Y, Greenberg HB, Kalica AR, Flores J, Kapikian AZ (1983a) Direct isolation in cell culture of human rotaviruses and their characterization into four serotypes. J Clin Microbiol 18:310-317

Wyatt RG, Kapikian AZ, Mebus CA (1983b) Induction of cross-reactive serum neutralizing antibody to human rotavirus in calves after *in utero* administration of bovine rotavirus. J Clin Microbiol 18:505-508

Yamaguchi H, Inouye S, Yamauchi M, Monishima T, Matsuno S, Isomura S, Suzuki S (1985) Anamnestic response in fecal IgA antibody production after rotaviral infection of infants. J Infect Dis 152:398-400

Yolken RH, Wyatt RG, Zissis G, Brandt CD, Rodiguez WJ, Kim HW, Parrtto RH, Urrutia JJ, Matya L, Greenberg HB, Kapikian AZ, Chanock RM (1987a) Epidemiology of human rotavirus type 1 and 2 as studied by enzyme-linked immunosorbent assay. N Engl J Med 299:1156-1161

Yolken RH, Wyatt RG, Mata L, Urrutia RH, van Kirk DH, Dolin R, Greenberg HB, Chanock RM (1978b) Secretory antibody directed against rotavirus in human milk: Measurement by means of enzyme linked immunosorbent assay. J Pediatr 93:916-921

Zissis G, Lambert JP, Marbehant P, Marissens D, Lobmann M, Charlier P, Delem A, Zygraich N (1983) Protection studies in colostrum-deprived piglets of a bovine rotavirus vaccine candidate using human rotavirus strains for challenge. J Infect Dis 148:1061-1068

Rabies Oral Immunization

Hilary Koprowski

The Wistar Institute, Thirty-Sixth Street at Spruce, Philadelphia, PA 19104-4268, USA

My interest in oral immunization dates to 1950 when I fed a child the live, attenuated polio virus vaccine. During the ensuing several years, many more children were immunized with this vaccine, and these efforts culminated in mass oral immunization of 250,000 infants in the Belgian Congo, during the epidemic of true infantile paralysis. This immunization required six weeks to perform. Our current interest in oral immunization stems from the fact that our real problem, as far as rabies is concerned, is not domestic animals, and not even vaccines for humans, but for effective vaccines for wildlife. The experience, particularly of Europeans with foxes, has indicated that the eradication of rabies in wildlife cannot be achieved by reducing the reservoir species by killing, gassing, or any other rather barbaric method. Instead, immunization of the wild reservoir animals offers an alternative approach (Steck et al. 1982). The questions are, how and what type of vaccine could be used, and how effective it would be - and this is the subject of my contribution. I will be discussing only one type of vaccine - the vaccinia-rabies glycoprotein recombinant vaccine (V-RG). I do not need to present details about the vaccine, other than that it is obtained by the cDNA derived from the mRNA for rabies glycoprotein incorporated into a construct plasmid from which it is in turn incorporated with vaccinia virus TK gene plasmid (Kieny et al. 1984). Transfection of vaccinia-infected cells with the plasmid allows double reciprocal recombination *in vivo* and final construction of the V-RG recombinant virus vaccine with the rabies glycoprotein gene interrupting the vaccinia TK gene sequence.

The first problem concerns the effectiveness of this vaccine. The original work done by my colleague, the late Dr. Wiktor, showed that one can immunize intradermally in the footpad, detect the virus neutralization antibodies (VNA), and demonstrate the protection following the challenge (Wiktor et al. 1984). Interestingly, V-RG can be inactivated by β-propriolactone and remains quite effective for immunization (Table 1). Thus, essentially, one does not need to have live vaccine in the preparation but using killed vaccinia would make the vaccine extremely costly, whereas live vaccinia vaccine is rather inexpensive.

In another approach of immunization with V-RG recombinant virus, a different species (rabbits) were injected intradermally which resulted in extremely high titers of anti-rabies antibodies (Wiktor et al. 1985). Actually, in my experience of working with rabies virus for so many years, I think the V-RG is the best vaccine I have ever encountered as far as immune response is concerned.

Now we come to the question of safety. Upon immunization by injection into the footpads, there was no trace of disease. When injected intraperitoneally, two of two mice showed lesions with the parental strain of vaccinia. As has been observed by many workers (Buller et al. 1985), the V-RG recombinant vaccine is less virulent than the parental strain (Table 2) (P. Desmettre, personal communication). Subsequent studies showed the results of immunization of animals by the oral route with this V-RG. Foxes have been orally immunized with V-RG (Blancou et al. 1986) and then

Current Topics in Microbiology and Immunology, Vol. 146

sacrificed to determine from which tissue the recombinant vaccine can be isolated. As can be seen in Table 3, the virus can be isolated only from tonsils for up to 48 hours and, after this time one cannot isolate the virus at all (J. Blancou, unpublished data).

Table 1. Induction of VNA in mice and protection from rabies

Animals/ inoculation route	Vaccine*	VNA titers					
		Rabies				Vaccinia	
		Day 0	Day 5	Day 11	Day 14	Day 14	Protection**
Mice/ intradermal	V-RGpro8***	<10	--	--	>30,000	250	12/12
	V-RGleu 8	<10	--	--	<10	--	0/12
	Vaccinia	<10	--	--	<10	250	0/12
Mice/ footpad	V-RGpro8	<10	--	--	>30,000	1250	12/12
	V-RGleu8	<10	--	--	<10	--	0/12
	Vaccinia	<10	--	--	<10	1250	0/12

* Vaccine was inoculated on day 0 using 2 x 10^8 pfu (intradermal) or 5 x 10^7 pfu (footpad) of virus.

** A challenge dose of 2400 or 24,000 mouse LD_{50} of MD5951 rabies virus was given on day 14 to mice and rabbits, respectively, by intracerebral inoculation.

***Difference of having leucine at position 8 in the rabies glycoprotein sequence.

Table 2. Safety in immunodepressed mice and swiss nude (Nu/Nu), females 6 wks old

1 - Inoculation in foot-pad

Doses (pfu)	Mice with reaction / Challenged mice 10^7	10^8
Parental strain "Copenhague"	0/2	0/2
Recombinant V-Rg	0/2	0/2

2 -Intraperitoneal route

Doses (pfu)	Mice with reaction / Challenged mice 10^7	10^8
Parental strain "Copenhague"	0/2	2/2 *
Recombinant V-Rg	0/2	0/2

* Cutaneous lesions on the tail with virus isolation (No isolation from the blood or the liver of these animals, sacrified 14 days after inoculation).

Conclusion: Total safety, both with respect to local and to general reactions.

(We are grateful to M.P. Kieny et al. for providing their as yet unpublished data which are presented in this table).

Table 3. Distribution of V-RG recombinant virus in foxes infected by the oral route

Seven foxes received 1 ml of V-RG (10^8 pfu) per os. They were euthanased after:
- 12 hours : 2 foxes
- 24 hours : 2 foxes
- 48 hours : 2 foxes
- 96 hours : 1 fox

- Different tissues obtained are assayed on Vero Cells for virus isolation (see table)
- These results will be completed with another technology : IF or tissue-culture.

Tissue	Hour - Post Immunization -			
	12	24	48	96
Buccal mucosa	-	-	-	-
Tonsil	+ (1/2)	+ (2/2)	+ (2/2)	-
Parotid lymph nodes	-	-	-	-
Submandibular lymph nodes	-	-	-	-
Mesenteric lymph nodes	-	-	-	-
Blood	-	-	-	-
Spleen	-	-	-	-
Brain	-	-	-	-
Salivary glands	-	-	-	-
Feces	-	-	-	-

(We are very grateful to J. Blancou et al. for providing their as yet unpublished data which is presented in this table).

In other studies, and this is really an experiment nearer to our problem, we have undertaken the oral immunization of raccoons (Rupprecht et al. 1986). I do not know whether you realize that in the Mid-Atlantic regions of the United States there are numerous rabid raccoons and the dangerous epizootic is spreading through larger and larger areas.

Studies of raccoons ingesting V-RG recombinant vaccine indicate that within 24 hours after vaccination the vaccinia recombinant can be isolated from buccal mucosa, parotid and submandibular lymph nodes and from tonsils. However, within the next 24 hours the virus cannot be isolated from any tissue of the raccoon (Table 4).

Turning from laboratory to the more practical aspects of immunization, Fig. 1 illustrates a bait which is used for oral immunization of raccoons (Rupprecht et al. 1987). It is a polyurethane foam sponge with the V-RG recombinant inside covered by wax, and soaked in fish oil, which raccoons find appealing. Incorporation of tetracycline into the bait allows determination, by fluorescence of their teeth, whether raccoons have actually swallowed the bait. The results allowed us to estimate that approximately 70% of animals ate the bait (this study was not done with rabies but was done with the bait itself).

Table 4. Distribution of a vaccinia-rabies glycoprotein (V-RG) recombinant virus in raccoons infected by the oral route.*

	Days post-immunization						
Tissue	1	2	3	5	9	11	14
Brainstem	-	-	-	-	-	-	-
Buccal mucosa	-	+	-	-	-	-	-
Tongue	-	-	-	-	-	-	-
Lung	-	-	-	-	-	-	-
Stomach	-	-	-	-	-	-	-
Tonsil	+	+	-	-	-	-	-
Esophagus	-	-	-	-	-	-	-
Salivary glands	-	-	-	-	-	-	-
Parotid/submandibular lymph nodes	+	+	-	-	-	-	-
Thymus	-	-	-	-	-	-	-
Trachea	-	-	-	-	-	-	-
Duodenum	-	-	-	-	-	-	-
Liver	-	-	-	-	-	-	-
Spleen	-	-	-	-	-	-	-
Kidney	-	-	-	-	-	-	-
Urinary bladder	-	-	-	-	-	-	-
Blood	-	-	-	-	-	-	-
CSF	-	-	-	-	-	-	-

* Following sedation, raccoons received 1.0 ml of V-RG ($10^{7.8}$ pfu/ml) per os and tissues obtained from 1-5 animals for viral isolation in BHK-21 cell cultures on the days indicated; + = V-RG isolated; - = no virus recovered.

Fig. 1. Bait courtesy of K. Lawson and D.H. Johnston

Table 5. Rabies virus-neutralizing antibodies and resistance to challenge

Vaccine & route	Dose (pfu)	Animals no.	Rabies neutralizing antibody titer*	Mean titer	Resistance to challenge**	Fraction surviving
No vaccine		#	0	NA	-	0/6
Conventional vaccine subcutaneous***		440	1.21	1.49	+	2/2
		441	1.77		+	
V-RG (intradermal)	10^8	446	3.03	2.82	+	2/2
		449	2.61		+	
V-RG (subcutaneous)	10^8	439	3.03	N/A	+	2/2
		442	0		+	
V-RG (oral scarified)	10^8	437	1.91	2.4	+	4/4
			438	2.33	+	
		447	2.61		+	
		452	2.75		+	
V-RG (oral)	10^4	408	1.35	N/A	+	
		427	0		-	1/4
		428	0		-	
		429	0		-	
	10^6	416	0.8	0.4	+	2/4
		425	0.8		+	
		426	0		-	
		430	0		-	
	10^8	414	2.33	2.57	+	4/4
		431	2.61		+	
		433	2.61		+	
		451	2.75		+	
V-RG (in bait)	10^8	411	1.07	1.8	- ^	4/5
		412	1.63		-	
		420	2.61		+	
		423	2.19		+	
		424	1.49		+ ^	

* Only the 28-day titer is given.

** +, resisted challenge; -, succumbed to rabies. All animals not resisting died between 15 and 25 days after challenge.

***Rabies in ND vaccine, lot 5523; neutralizing antibody titers obtained with live attenuated virus are similar (not shown).

^ Two animals were observed to have ingested only a part of the vaccine.

(We wish to thank J. Blancou et al. for providing their published data which are presented in this table).

When raccoons were vaccinated with V-RG by various routes, including ingestion of V-RG bait, and then challenged with virulent virus (Table 5) 10^8 pfu of V-RG virus in bait immunized as effectively as by any other route of vaccination (Rupprecht et al. 1986). In another experiment raccoons were immunized with V-RG bait and the level of neutralizing antibodies was measured before and after challenge with virulent virus (Rupprechet al. 1986) (Table 6).

Table 6. Rabies virus-neutralizing antibodies (VNA) and protection from rabies challenge in raccoons receiving an oral vaccinia-rabies glycoprotein recombinant (V-RG) virus vaccine

Experimental treatment method	Dose	Animal #	VNA titer (IU/ml)* PI	PC (Day 30)	PC (Day 90)	Response to challenge+
V-RG in sponge bait	10^8 pfu/ml	49	54.0	464.0	18.0	S
		48	18.0	163.0	12.0	S
		62	18.0	18.0	18.0	S
		58	6.0	18.0	18.0	S
		59	6.0	18.0	12.0	S
		63	2.0	18.0	18.0	S
		56	0.6	163.0	18.0	S
		57	0.6	18.0	6.0	S
V-RG by oral	10^6 pfu/ml	40	1.3	0.7**	--	D(13)
infusion	10^8 pfu/ml	41	6.0	9.9	36.0	S
Cell culture media in sponge bait		1	<0.3	2.0**	--	D(17)
		2	<0.3	2.0**	--	D(10)
		3	<0.3	0.7**	--	D(21)
		4	<0.3	2.0**	--	D(12)
		5	<0.3	0.7**	--	D(10)
		6	<0.3	1.0**	--	D(17)

* VNA were determined 28 days post-immunization (PI)(Day 0 of challenge), and 30 and 90 days post-challenge (PC) or **on the day of death. All experimental animals were seronegative for rabies (<0.3 IU) prior to immunization.

+ Animals were challenged intramuscularly 28 days PI with 0.3 ml ($10^{5.5}$ $MICLD_{50}$) of rabies virus strain MD5951; S = Survived, D(n) = Died (day of death PC).

The problem of rabies in dogs is of importance in countries, such as India, where endemic rabies is spread by stray dogs. In India, there are between 20,000 human cases of rabies per year, and the wildlife are not a problem. Instead, there is the problem of stray dogs which cannot be controlled by any other means except immunization. Thus, oral immunization for stray dogs in such countries with the vaccinia-rabies recombinant is indicated. As shown in Table 7, in order to immunize dogs effectively by the oral route, one has to use rather large doses of V-RG virus (P. Desmettre, unpublished data) when compared to those used for raccoons or foxes. The results also indicate that if unvaccinated dogs are kept in close contact with vaccinated animals there is no transmission of the infection.

Table 7. Oral vaccination of dogs with the V-RG recombinant virus vaccine

		Rabies antibodies (RFFIT)							
	No. of	Day 0		Day 7		Day 14		Day 28	
Group	Dogs	log	IU	log	IU	log	IU	log	IU
Vaccinates	259	Neg	Neg	Neg	Neg	Neg	Neg	Neg	Neg
$10^{8.6}$ pfu	403	Neg	Neg	1.4	1.2	1.9	3.6	2.4	12.6
	291	Neg	Neg	Neg	Neg	Neg	Neg	Neg	Neg
	278	Neg	Neg	Neg	Neg	Neg	Neg	Neg	Neg
Unvaccinated controls, in contact with the $10^{8.6}$ pfu vaccinates	402	Neg	Neg	Neg	Neg	Neg	Neg	Neg	Neg
	296	Neg	Neg	Neg	Neg	Neg	Neg	Neg	Neg
Vaccinates	280	Neg	Neg	1.7	2.5	2.2	7.9	2.1	6.1
$10^{9.6}$ pfu	298	Neg	Neg	2.0	5.1	2.5	13.7	2.5	13.3
	404	Neg	Neg	1.1	0.5	1.8	3.0	1.9	3.6
	281	Neg	Neg	1.1	0.6	1.6	1.9	1.9	3.6
Unvaccinated	275	Neg	Neg					Neg	Neg
controls	290	Neg	Neg					Neg	Neg
	283	Neg	Neg					Neg	Neg
	279	Neg	Neg					Neg	Neg
	277	Neg	Neg					Neg	Neg

(We are very grateful to P. Desmettre et al. for providing their as yet unpublished data which are presented in this table).

Cats are of importance in spreading rabies in urban areas. As shown in Table 6, they can be effectively immunized against rabies with V-RG recombinant vaccine by the oral route (P. Desmettre, unpublished data), with an effective immunizing dose lower than that necessary for dogs (Table 8).

In Europe, foxes are the main species infected with rabies virus. Infected foxes were first seen after World War II in Eastern Europe and the Soviet Union. The unsuccessful programs of eradication by gassing was replaced in this decade by oral immunization. The late Dr. Steck, in Switzerland, initiated this program. Good results have been obtained after immunization with the V-RG recombinant vaccine (Blancou et al. 1986) (Table 9). Immunizing doses of between 10^6 and 10^7 plaque-forming units are effective against challenge with lethal virus. In unvaccinated controls, the same challenge dose caused rabies, and the animals died. The kinetics of immune response for a 12 month period are presented in Table 10 (Blancou, unpublished data). Foxes were vaccinated with 10^8 plaque-forming units given orally. Neutralizing antibodies were detected through 350 days after immunization except for a few foxes which showed relatively low levels of neutralizing antibodies, although they were all immune (Blancou et al.

Table 8. Oral vaccination of cats with the V-RG recombinant virus vaccine

Group	No. of cats	Rabies antibodies (RFFIT) Day 4 Log	Day 4 IU	Day 10 Log	Day 10 IU	Day 21 Log	Day 21 IU	Day 28 Log	Day 28 IU	Results after intramuscular challenge on day 52 *
Unvaccinated controls	838	Neg	Neg	Neg	Neg	Neg	Neg	Neg	Neg	D (22)
	839	Neg	Neg	Neg	Neg	Neg	Neg	Neg	Neg	D (18)
	843	Neg	Neg	Neg	Neg	Neg	Neg	Neg	Neg	D (12)
	844	Neg	Neg	Neg	Neg	Neg	Neg	Neg	Neg	D (12)
	847	Neg	Neg	Neg	Neg	Neg	Neg	Neg	Neg	D (16)
Vaccinates 10^4 pfu	829	Neg	Neg	Neg	Neg	Neg	Neg	Neg	Neg	D (12)
	836	Neg	Neg	Neg	Neg	Neg	Neg	Neg	Neg	D (19)
	849	Neg	Neg	Neg	Neg	Neg	Neg	Neg	Neg	D (15)
Vaccinates 10^6 pfu	840	Neg	Neg	2.6	17.4	2.7	22.0	2.5	14.2	S
	842	Neg	Neg	2.5	12.6	2.4	19.0	2.4	10.0	S
	848	Neg	Neg	Neg	Neg	Neg	Neg	Neg	Neg	S
Vaccinates 10^8 pfu	834	Neg	Neg	2.4	10.7	2.0	8.9	2.2	7.8	S
	841	Neg	Neg	2.8	27.0	2.5	14.2	2.4	11.5	S
	846	Neg	Neg	2.5	14.2	2.5	14.2	2.4	10.0	S

*D = died
S = survived
(13) = time before death following challenge

(We are very grateful to P. Desmettre et al. for providing their as yet unpublished data which are presented in this table).

Table 9. Oral vaccination of foxes with the V-RG recombinant virus vaccine, test N. 3 : Effective dose

Group	No. of fox	Rabies antibodies ** D14	D28	Resistance to challenge *
Vaccinates 10^6 pfu	539	0.08	Neg	D (15)
	547	Neg	Neg	D (16)
	552	Neg	Neg	S
	556	2.31	1.68	S
Vaccinates 10^7 pfu	542	13.48	11.56	S
	543	1.23	24.65	S
	548	Neg	0.15	S
	550	0.96	0.25	S
Vaccinates 10^8 pfu	517	0.46	0.09	S
	534	6.07	6.96	S
	541	0.08	0.95	S
	537	1.93	6.07	S
Unvaccinated controls in contact with the group vaccinated with 10^8 pfu	523***	0.09	0.04	S
	535	Neg	Neg	D (17)
	536	Neg	Neg	D (17)
	545	Neg	Neg	D (15)
Unvaccinated controls	501	Neg	Neg	D (18)
	553	Neg	Neg	D (17)

*Intramuscular challenge at $10^{4.7}$LD50/IM/Fox ($10^{4.2}$LD50/IC/Mice)
S = survived
D = died
(17) = time before death following challenge,
**IU/ml (RFFIT)
***Fox n. 523 vaccinated by accidental biting

1988). Apparently, other immune mechanisms that protect an animal against rabies in the absence of neutralizing antibodies, are involved in resistance to challenge.

Fox cubs with very sharp teeth are often the best transmitters of rabies in Europe and in Canada. Fox cubs were vaccinated (Brochier et al. 1988) and then challenged 365 days later. Neutralizing antibodies were measured after a single immunization per cub (Table 11). The immunity lasted for at least 365 days (at the time of last testing) and probably lasts much longer.

Skunks are animals which spread rabies both in the United States and in Canada. Again, the only effective way to control rabies in skunks is through oral immunization. The results obtained in skunks by oral immunization (Tolson et al. 1987), compared to scarification and direct intestinal placement of a capsule, are presented in Table 12. The results are expressed in virus-neutralizing antibodies and the response is perhaps lower than after parenteral, particularly intramuscular, immunization. However, the animals which have no measurable level of neutralizing antibodies may also be resistant to challenge.

Table 10. Oral vaccination of foxes with the V-RG recombinantvirus vaccine, test N. 1 : Rabies antibody kinetics and immunity over 12 months

Group	No of fox	Rabies Antibodies (IU) D 0	D 30	D 90	D 180	D 350	Resistance to challenge on D 350
Vaccinates	401	Neg	0.87	0.04	0.03	Neg	S
(10^8 pfu)	402	Neg	2.49	0.46	0.34	0.08	S
	404	Neg	0.46	0.03	0.01	Neg	S
	406	Neg	24.6	1.12	0.77	0.19	S
Controls	560						D (20)
	561						D (21)
	563						D (20)

S = survived
D = died
(20) = time before death following challenge

(We wish to thank J. Blancou et al. for providing their published data which are presented in this table).

The experiment with ferrets (Table 13) was performed principally because of interest in contact control: ferrets bite each other when housed in the same cage. The animals were vaccinated orally with the dosage of 10^8-10^9 plaque-forming units, and then kept together with the contact controls. The contact controls were perfectly susceptible to infection: they had no neutralizing antibodies whereas there was a strong serum-conversion in ferrets orally immunized with vaccinia recombinant (P. Desmettre, unpublished data).

The problem of susceptibility of various animal species to live vaccinia was addressed by the French and Belgian colleagues (Brochier et al. 1988) who have used crows and tried to find if birds which may prey on carcasses of other animals are immunized and possibly may excrete vaccinia. The results (Table 14) show that only a few birds died with intercurrent infection while the rest survived the imunization with V-RG recombinant vaccine.

Finally, the last species which is of importance in the possible transmission of rabies are badgers. These results indicate that 10^8 plaque-forming units is not a sufficient amount to immunize all badgers and they are probably similar to dogs in which doses of 10^9-$10^{9.6}$ plaque-forming units must be used for immunization. With such dosages the animals are likely to be immunized (Brochier et al. 1988).

Table 11. Oral vaccination of fox cubs with the V-RG virus vaccine, recombinant test N. 6 : Immunity persistence (vaccination at $10^{7.2}$ pfu)

Group	No. of fox cubs	Weight (KG) on day 0	Rabies antibodies Reffit (UI/al) D 3	Reffit (UI/al) D 14	Reffit (UI/al) D 28	SN (UI/al) D 28	Resistance to challenge D 33 *	Resistance to challenge D 180 *	Resistance to challenge D 360 *
	152	2.0	Neg	4.45	9.95	9.97	S		
Vaccinates	153	1.8	Neg	3.07	4.92	8.19	S		
challenged	154	1.8	Neg	7.85	20.9	>22.25	S		
on day 33	155	2.0	Neg	1.57	9.95	2.44	S		
	156	2.0	Neg	1.37	3.06	1.16		S	
Vaccinates	157	2.0	Neg	3.06	5.25	5.82		S	
challenged	158	1.8	Neg	5.25	4.45	4.45		S	
on day 180	159	2.2	Neg	3.06	5.25	5.82		S	
Vaccinates	160	2.8	Neg	1.52	4.45	2.16			**
challenged	161	2.4	Neg	1.82	2.81	0.82			S
on day 365	175	1.3	Neg	0.36	2.81	0.67			S
(June 87)	176	1.4	Neg	<0.178	<0.178	< 0.30			D
Vaccinates	172	2.3	Neg	<0.458	0.89	<0.30	S		
Unvaccinated controls in	173	2.2	Neg	<0.458	<0.458	<0.30	D (26)		
contact with N. 172	174	2.2	Neg	<0.458	<0.458	<0.30	D (25)		
Unvaccinated	150	3.0	Neg	<0.233	<0.390	<0.30	D (23)		
controls on day 33	151	3.4	Neg	<0.233	<0.390	<0.30	D (25)		

(*) challenge performed with $10^{3.2}$ LD_{50}/mice (1 ml in the temporal muscle)
S = survived, D = died, (23) = time befoe death following challenge, ** = accidentally died on day 361

(We wish to thank B. Brochier et al. for providing their published data which are presented in this table)

Table 12. Vaccination of skunks with V-RG recombinant virus vaccine

Vaccination route and skunk number	Serum neutralizing antibody titers (IU)				
	Day 0	Day 14	Day 28	Day 60	Day 90
Oral					
1	0	0.41	2.52	2.35	0.85
2	0	0	0.17	0.22	0
3R	0	0	1.43	0.35	0.17
4	0	0	2.43	0.65	0.24
5D	0	0	0	0	0
6	0	0	4.61	4.59	1.41
7	0	0	0.78	1.20	0.39
8R	0	0	0	0	0
Intestinal					
1R	0	0	0	0	0
2R	0	0	0	0	0
3	0	3.48	1.35	3.78	0.96
4	0	2.13	2.07	1.04	0.28
5R	0	0	0	0	0
6	0	14.4	10.6	3.76	1.70
7	0	1.59	0.87	0.80	0.52
8R	0	2.74	0.85	0.65	0.52
Intramuscular					
1R	0	22.3	12.5	1.20	0.37
2	0	21.4	14.4	8.37	3.78
3	0	37.4	20.4	2.61	2.87
Scarification					
1	0	156	34.3	4.57	2.61
2R	0	44.5	6.78	1.93	0.35
3	0	159	15.7	2.57	1.39
4	0	68.5	5.35	2.30	0.70
5	0	101	11.1	2/37	0.70
6	0	48.4	11.1	2.65	0.70

R - succumbed to rabies on challenge; FA positive
D - death not due to rabies; FA negative

(We wish to thank N.D. Tolson et al. for providing their published data which are presented in this table).

Table 13. Safety and activity in ferrets

- Safety of the recombinant was demonstrated in 4 ferrets vaccinated with 1 ml of virus suspension (10^8 or 10^9 pfu/ml) by direct oral route.

- No diffusibility was brought to light on contact controls (in contact with 10^9 pfu vaccinated animals).

- Seroconvertion is observed in the two ferrets which received 10^9 pfu, and in one of the two which received 10^8 pfu.

Group	No. of ferret	Rabies antibodies (IV)	
		D 0	D 28
Vaccinates	1 a	Neg	Neg
10^8 pfu	1 b	Neg	15.3
Vaccinates	2	Neg	3.2
10^9 pfu	4	Neg	3.2
Contact-controls	3	Neg	Neg
	5	Neg	Neg
Controls	6 a	Neg	Neg
	6 b	Neg	Neg

(We are grateful to P. Desmettre et al. for providing their as yet unpublished data which are presented in this table).

In summary, today we are in a situation where we know that vast numbers of species can be effectively immunized by oral feeding with V-RG recombinant virus vaccine. It is probably the most powerful vaccine ever developed against rabies. In the field tests conducted in Belgium (Pastoret et al. 1988), foxes were immunized in the wild with V-RG in bait and are being allowed free contact with other species to find out at the end of a year or two, whether there is transmission of the vaccine. In the United States, plans are being developed for oral immunization of raccoons on islands off the Virginia and South Carolina shorelines. Hopefully during the year 1988, it will be possible to evaluate results of these restricted field trials and then to move on to stamp out the rabies plague in raccoons by effective oral immunization of the animals.

It is also hoped that oral immunization of stray dogs in such rabies-endemic countries as Egypt and India may break the link of rabies transmission and lead ultimately to an effective control of rabies in the world.

Table 14. Safety in crows

- Sixteen black crows received 1 ml (10^8 pfu) of V-RG recombinant by oral route. Five crows were kept as controls.
- A few days after inoculation, 5 animals (3 vaccinates and 2 controls) died, obviously because of the containment conditions.
- During a 28 days post-inoculation period, all the animals remained healthy without any poxivirus lesion.

All animals were euthanased.

Group	No. of crows	Observations *
	1	D (3)
	2	S
	3	S
Vaccinates	4	S
	5	S
10^8 pfu	6	S
	8	S
	9	S
	10	S
	11	S
	12	S
	13	D (4)
	24	D (5)
	25	S
	17	D (3)
Controls	18	S
	19	S
	20	D (3)
	23	S

*D = died
(4) = no. of days after inoculation
S = survived without any lesion

(We wish to thank B. Brochier et al. for providing their published data which are presented in this table).

REFERENCES

Blancou J, Kieny M-P, Lathe R, Lecocq JP, Pastoret RP, Soulebot JP, Desmettre P (1986) Oral vaccination of the fox against rabies using a live recombinant vaccinia virus. Nature 332:373-375

Blancou J, Pastoret PP, Brochier B, Kieny MP, Desmettre P, Artois M (1988) Vaccination des renards contre la rage avec le virus recombinant vaccine-rage. Comm for Commun p 112

Brocheir B, Languet B, Blancou J, Thomas I, Kieny MP, Lecocq JP, Desmettre P, Pastoret PP (1988) Innocuite du virus recombinant vaccinia-rage chez quelques especes non-cibles. Comm for Commun p 118

Buller RML, Smith GL, Cremer K, Notkins AL, Moss B (1985) Decreased virulence of recombinant vaccinia virus expression vectors is associated with a thymidine kinase-negative phenotype. Nature 317:813-815

Kieny M-P, Loathe R, Drillien R, Spehner D, Skory S, Schmitt D, Wiktor TJ, Koprowski H, Lecocq J-P (1984) Expression of rabies virus glycoprotein from a recombinant vaccinia virus. Nature 312:163-166

Pastoret PP, Brochier B, Paquot A, Costy F, Kieny MP, Lecocq JP, Languet B, Desmettre P, Antonire H, De Bruyn J (1988) Premier essai du virus recombinant vaccine-rage sur la terrain. Comm for Commun p 124

Rupprechet CE, Wiktor TJ, Hamir AN, Dietzschold B, Wunner WH, Glickman LT, Koprowski H (1986) Oral immunization and protection of raccoons (*Procyon lotor*) with a vaccinia rabies glycoprotein recombinant virus vaccine. Proc Natl Acad Sci USA 83:7947

Rupprecht CE, Hamir AN, Johnston DH, Koprowski H (in press) Efficacy of a vaccinia-rabies glycoprotein recombinant virus vaccine in raccoons (*Procyon lotor*). Rev Inf Dis

Rupprecht CE, Johnston DH, Hamir A, Dietzschold B, Koprowski H (1987) Preliminary studies on the immunization and field baiting trials of raccoons against rabies. Cold Spring Harbor Laboratory, New York, p 389

Steck FA, Wandeler A, Bischel B, Capt S, Schneider L (1982) Oral immunization of foxes against rabies. Abl Vet Met 29:372

Tolson ND, Charlton KM, Stewart RB, Campbell JB, Wiktor TJ (1987) Immune response in skunks to a vaccinia virus recombinant expressing the rabies virus glycoprotein. Can J Vet Res 51:363-366

Wiktor TJ, MacFarlan RI, Reagan KJ, Dietzschold B, Curtis PJ, Wunner WH, Kieny M-P, Lathe R, Lecocq J-P, Mackett M, Moss B, Koprowski H (1984) Protection from rabies by a vaccinia virus recombinant virus containing the rabies glycoprotein gene. Proc Natl Acad Sci USA 81:7194-7198

Wiktor TJ, MacFarlan RI, Dietzschold B, Rupprechet CE, Wunner WH (1985) Immunogenic properties of vaccinia recombinant virus expressing the rabies glycoprotein. Ann Inst Pasteur/Virol 136E:405-411

Bacterial Vaccines

Induction of the Mucosal Immune Response

U. Dahlgren, B. Carlsson, F. Jalil, R. MacDonald, F. Mascart-Lemone, K. Nilsson, D. Roberton, F. Sennhauser, A. Wold, and L.Å. Hanson

Department of Clinical Immunology, University of Göteborg, Göteborg, Sweden; Department of Social and Preventive Pediatrics, King Edward Medical College, Lahore, Pakistan; Department of Immunology, Hospital Saint-Pierre, Free University of Brussels, Belgium; and Department of Immunology, Royal Children's Hospital,

INTRODUCTION

Exposure to antigens via mucosal membranes is well known to induce responses in the common mucosal system dominated by secretory IgA (SIgA), whereas parenterally given antigens are generally not considered to induce SIgA antibody production. Recent studies on the effect of vaccination suggests, however, that the antibody response patterns may be rather more complex, with many exceptions from the above concept (Mestecky 1987). Thus, parenteral vaccination can lead to a mucosal response, while intestinal exposure may result in predominantly serum responsiveness. This brief review will bring out some recent studies illustrating the wide range of conditions influencing the mucosal antibody response.

PARENTERAL IMMUNIZATION AND ITS INFLUENCE ON THE SIgA RESPONSE

Using parenteral inactivated poliovirus for vaccination in humans, we could show that after 2-3 doses, SIgA antibodies against poliovirus were detectable in saliva (Mellander et al. 1985). In a recent study by Levenson et al. (1988) a *Shigella* vaccine induced brisk responses of SIgA to the *Shigella* O antigen in the saliva of parenterally immunized monkeys, as well as of serum IgM, IgA, and IgG. This was already obtained after the first dose and a booster resulted in antibody responses in all classes when given either as a second injection or by oral challenge.

Mucosal priming, followed by parenteral boosting, may induce an efficient mucosal IgA response. Thus, we have shown that parenteral immunization with cholera and poliovirus vaccines boost not only serum antibody responses but also salivary and milk SIgA antibodies in individuals who have been previously exposed naturally (Svennerholm et al. 1980; Svennerholm et al. 1981).

In a recent extension of this work, we found that immunization against cholera in naturally exposed mothers induced a serum IgA antibody response to the *V. cholerae* 0 antigen, consisting of polymers, in 82 $\pm$ 7% (Mascart-Lemone et al. 1988). These IgA antibodies were mainly dimeric and did not contain secretory component. These polymeric IgA antibodies disappeared from serum more rapidly than the monomeric IgA antibodies. The SIgA response in these mothers' milk paralleled serum dimeric IgA antibody response.

Current Topics in Microbiology and Immunology, Vol. 146

Peroral vaccination with a live oral *Salmonella typhi* 21a vaccine also gave predominantly polymeric serum IgA production (72 ± 17%). However, the milk and serum IgA responses to the *S. typhi* 0 antigen were dissociated (Mascart-Lemone et al. 1988). It is not clear if the more rapid disappearance of the polymeric rather than monomeric IgA from serum could be due to transfer into secretions; it is more likely due to a shorter half-life of the polymers (Delacroix et al. 1983). The origin of the polymeric IgA in serum is not clear, but could be related to the large proportion of circulating IgA-committed B lymphocytes that produce polymeric IgA and presumably are on their way to home to mucosal tissues or exocrine glands (Kutteh et al. 1978). The production of monomeric or polymeric IgA might be related to the stage of maturation of the lymphocytes according to studies by Moldoveanu et al. (1984).

The dissociation of the serum and milk IgA antibody responses to *S. typhi* reviewed above does not support a mucosal origin of all the dimeric serum IgA as does the parallel serum and milk IgA response to *V. cholerae* (Mascart-Lemone et al. 1988). A study by Kentala et al. (1986) actually showed that after peroral immunization with *S. typhi*, IgA secreting cells appeared transiently in the circulation in the absence of serum and secretory IgA antibodies. This is in agreement with an earlier investigation where lactating mothers were colonized perorally with an *Escherichia coli* 083 strain which induced cells to appear in the milk which produced IgA anti-083 antibodies, although there were no such antibodies in the circulation at that time (Goldblum et al. 1975). It may be a matter of antigen dose as to whether or not antibodies appear free in the serum as well and not only in the secretion.

THE AVIDITY OF SIgA ANTIBODIES

Not only the level but also the avidity are important for the functional capacity of antibodies (Aslstedt et al. 1974). The avidity of SIgA antibodies in human milk has been investigated recently using a thiocyanate elution technique in an enzyme-linked immunosorbent assay (Roberton et al. 1988). It was observed that milk SIgA antibodies to *E. coli* 0 antigens were generally of higher avidity in Swedish than in Pakistani mothers. The latter were not undernourished, but were more heavily exposed to microbes than the Swedish women. This does not explain the lower antibody avidity and does not make obvious why the avidity increased in the early part of lactation, and again at a later time, decreased. The Swedish mothers maintained a similar median avidity during the period of lactation (Table 1). A mature pattern of antibody avidity was apparent in milk from Swedish mothers even on day three of lactation. The median avidity of milk SIgA antibodies to diphtheria toxoid were also higher in the Swedish than the Pakistani mothers (Table 1). The latter had only been naturally exposed and the former had been vaccinated in childhood. It is possible that the vaccine provides a larger antigen dose than does the infection.

Lactating Pakistani mothers who were vaccinated parenterally against cholera did not exhibit changes in the avidity of their milk SIgA antibodies to *V. cholerae* 0 antigen following immunization and the median avidity did not differ from that of unvaccinated mothers (Table 2). The pre-existing immune status of Pakistani mothers may explain, in part, this observation. The increased antibody titres in the milk sample from the vaccinated mothers were moderate. A low avidity of milk SIgA might be advantageous by furnishing the individual with antibodies that would react with an array of antigens.

Table 1. Avidity of milk secretory IgA antibodies

Origin of samples	Number of mothers	Time of lactation	Avidity[a] of antibodies to E. coli O antigens	Avidity[a] of antibodies to Diphtheria toxoid
Pakistan	12	17.5[b] (14-24)[c]	1.78 (0.60-3.35)	
			$p<0.02$	
Sweden	11	21	2.65 (1.85-3.05)	
Pakistan	12	50 (42-70)	2.50 (1.15-3.65)	
			n.s.	
Sweden	11	73 (35-105)	3.10 (1.70-5.0)	
Pakistan	12	mid-lactation		2.35(0.65-4.00)
				$p<0.002$
Sweden	14	mid-lactation		4.30 (2.20-5.0)

[a] Avidity expressed as molarity of thiocyanate used for elution in ELISA.
[b] Mean in days.
[c] Range.

Table 2. Avidity of milk SIgA antibodies in mothers vaccinated against cholera

	Number	Avidity[a] mean and (range) of anti-V. cholerae O antigen: Prevaccination	Two weeks post	Sixteen weeks post
Pakistani mothers	10	2.40 (1.64-3.47)	2.38 (1.94->5)	2.19 (1.25-4.42)
Australian mothers	8	2.68 (1.59-3.66)		

[a] Avidity is expressed in molarity of thiocyanate used for elution in ELISA.

IgM AND IgG ARE PART OF THE MUCOSAL IMMUNE SYSTEM

Some studies have suggested that not only SIgA, but also IgM and IgG in secretions may, at least in part, be of local origin. Thus a difference in electrophoretic mobility was noted for IgG in milk when compared with serum (Hanson 1961), and Keller et al. (1983) found evidence that the mammary gland produces the IgG4 isotype. Rats immunized in the Peyer's patches with *E. coli* had IgA as well as IgG and IgM antibodies antibodies in milk and bile against the LPS and pili antigens. In milk and in bile the IgG and IgM antibody levels, when compared with serum, differed between the anti-pili and the anti-LPS antibodies and would indicate that IgG and IgM were produced locally in the mammary gland and in the liver (Dahlgren et al. 1987). In addition intravenous injection to rats of IgG and IgM antibodies against ferritin did not result in any IgM and only traces of IgG antibodies in the bile and none in milk. The production of IgM in the liver has been confirmed by other studies as well (Jackson et al. 1985; Dahlgren et al. 1986; Nilsson et al. 1988a). Experiments performed in germfree rats monocolonized with *E. coli* support these findings (Dahlgren et al. 1988).

Cell transfer studies using mesenteric lymph node cells from rats immunized in the Peyer's patches verified the homing of IgM and IgG immunocytes to mucosal sites or exocrine glands, just as for IgA producing cells (Dahlgren et al. 1987). Previous studies have not shown any transfer of IgA from serum to the milk in rats or mice (Dahlgren et al. 1981; Koertge and Butler 1986), which is in disagreement with Halsey et al. (1983).

THE ANTIGEN DEPENDENCE OF LYMPHOCYTE HOMING TO MUCOSAL SITES

In previous studies, we have shown that rats immunized in the Peyer's patches with type 1 piliated *E. coli* 06 elicited IgA anti-06 antibodies mainly in the bile and IgA anti-pili antibodies predominantly in the milk. This was confirmed by cell transfer studies (Dahlgren et al. 1987). When lactating rat dams were subjected to lymph drainage it was found that the biliary IgA anti-pili antibodies decreased less than did the IgA anti-LPS antibodies. There was no apparent difference in antibody titers in non-lactating rats. This suggests that a greater proportion of the IgA anti-pili than anti-06 antibodies originate from extraintestinal sites in the lactating rat. A portion of the biliary IgA anti-pili antibodies presumably originates in the mammary gland, or in the liver. Transfer of mesenteric lymph node cells from immunized donors to non-lactating recipients resulted in the appearance of both IgA anti-pili and IgA anti-06 antibodies in the bile (Nilsson et al. 1988b). In lactating rats, the IgA anti-pili antibodies appeared almost exclusively in the milk, and not in the bile after cell transfer (Dahlgren et al. 1987); these results might indicate that lactogenic hormones are involved in the regulation of cell traffic. There is some evidence that homing of IgG and IgM immunocytes may also be influenced by their antigen specificity in a similar manner (Dahlgren et al. 1988). These studies suggest that it may be useful to test whether vaccine candidates which induce mucosal immunity really do so at sites where responses are desired.

DIFFERENCES IN SIgA ANTIBODY RESPONSES TO VARIOUS ANTIGENS

There are several studies to suggest that more efficient SIgA responses may occur to live than nonreplicating antigens (Pierce et al. 1988). We have observed that it was difficult to induce mucosal responses to ovalbumin when given perorally to rats. In fact, direct immunization of the Peyer's patches with this antigen failed to induce IgA anti-ovalbumin responses in bile or in milk, however, IgG responses were seen (Wold et al. 1987). This poor response to a food protein was quite different from that obtained against *E. coli* used in the experiments described above. We therefore compared the effects of live type 1 piliated *E. coli* 06 monocolonizing germ-free rats fed ovalbumin and β-lactoglobulin. Antibodies of the IgG, IgM and IgA isotypes against 06 appeared within one week after colonization in both serum and bile (Wold et al. 1988). However, the antibody response to the type 1 pili was detected a few weeks later; this response included IgG as well as IgM and IgA antibodies. The response to the food proteins was low and did not appear until after 16 weeks. Both IgA and IgM were found in milk against ovalbumin; however, only IgA β-lactoglobulin responses were seen, and many animals did not respond at all to these proteins. Antibody-forming cells as determined with the ELISPOT assay were few against the food proteins and about 10 times as common against the bacterial antigens. Food protein-reactive cells were somewhat more common in the mesenteric lymph nodes than in the spleen. The reverse was true for the cells directed against the bacteria. The milk, however, did not show dramatic differences between antibodies to bacterial or food antigens, as did serum and bile. This might indicate that the homing of lymphoblasts to the lactating mammary gland may not parallel the homing of cells to other mucosal sites. By analogy, a study of the response after oral cholera vaccination showed no correlation between antibody titers in intestinal lavage and milk, whereas salivary titers were correlated (Jertborn et al. 1985).

These observations illustrate rather strikingly how food and bacterial antigens exposing mucosal membranes are handled differently. The mechanisms behind this are still obscure.

Acknowledgements. These studies were supported by grants from the Swedish Medical Research Council (No 215); the Swedish Agency for Research Cooperation with Developing Countries; The Ellen, Walter and Lennart Hesselman Foundation for Research; The Fond National de la Recherche Scientifique, Brussels, Belgium; and WHO.

REFERENCES

Ahlstedt S, Holmgren J, Hanson LÅ (1974) The protective capacity of antibodies against *E. coli* 0 antigen with special reference to the avidity. Int Archs Allergy Appl Immnol 46:470-480

Carlsson B, Zaman S, Mellander L, Jalil F, Hanson LÅ (1985) Secretory and serum immunoglobulin class-specific antibodies to poliovirus after vaccination. J Infect Dis 152:1238-1244

Dahlgren U, Ahlstedt S, Hanson LÅ (1986) Hepatic production of biliary IgM. Scand J Immunol 24:615-619

Dahlgren U, Ahlstedt S. Hanson LÅ (1987) The localization of the antibody response in milk or bile depends on the nature of the antigen. J Immunol 138:1397-1402

Dahlgren U, Ahlstedt S, Hedman L, Wadsworth C, Hanson LÅ (1981) Dimeric IgA in the rat is transferred from serum into bile but not into milk. Scand J Immunol 14:95-98

Dahlgren UIH, Wold AE, Hanson LÅ, Midtvedt T (1988) Antibody response in germ-free rats monocolonized with *E. coli*. II. Difference in the localization to various secretions of anti-LPS and anti-fimbrial antibodies. (Submitted for publication)

Delacroix DL, Elkon KB, Geubel AP, Hodgson HF, Dive C, Vaerman JP (1983) Changes in size, subclass, and metabolic properties of serum immunoglobulin A in liver diseases with high serum immunoglobulin A. J Clin Invest 71:358-367

Goldblum RM, Ahlstedt S, Carlsson B, Hanson LÅ, Jodal U, Lidin-Janson G, Sohl-Åkerlund A (1975) Antibody-forming cells in human colostrum after oral immunization. Nature 257:797-799

Halsey JF, Mitchell CS, McKenzie SJ (1983) The origin of secretory IgA in milk: A shift during lactation from a serum origin to local synthesis in the mammary gland. Ann NY Acad Sci 409:452-459

Hanson LÅ (1961) Comparative studies of the immune globulins of human milk and of blood serum. Int Archs Allergy 18:241-267

Jackson GDF, Lemaitre-Coelho I, Vaerman J-P, Bazin H, Beckers A (1985) A role for the spleen in the appearance of IgM in the bile of rats injected intravenously with horse erythrocytes. J Immunol 135:152-157

Jones CL, MacDonald RA, Hosking CS, Roberton DM (1987) Estimating the relative avidity of mucosal IgA for antigen. J Immunol Methods 105:111-117

Keller MA, Heiner DC, Kidd RM, Myers AS (1981) Local production of IgG4 in human colostrum. J Immunol 130:1654-1657

Koertge TE, Butler JE (1986) Dimeric M315 is transported into mouse and rat milk in a degraded form. Mol Immunol 23:839-845

Kutteh WH, Koopman WJ, Conley ME, Egan ML, Mestecky J (1980) Production of predominantly polymeric IgA by human peripheral blood lymphocytes stimulated *in vitro* with mitogens. J Exp Med 152:1424-1431

Levenson VI, Chernokhvostova EV, Lyubinskaya MM, Salamatova SA, Dzhikidze FK, Stasilevititch ZK (1988) Parenteral immunization with Shigella ribosomal vaccine elicits local IgA response and primes for mucosal memory. Int Archs Allergy Appl Immunol (In press)

MacDonald RA, Hosking CS, Jones CL (1987) The measurement of relative affinity by ELISA using thiocyanate elution. J Immunol Methods 106:191-194

Mascart-Lemone F, Carlsson B, Jalil F, Hahn-Zoric M, Duchateau J, Hanson LÅ (1988) Polymeric and monomeric IgA response in serum and milk after parenteral cholera and oral typhoid vaccination (Submitted for publication)

Mestecky J (1987) The common mucosal immune system and current strategies for induction of immune responses in external secretions. J Clin Immunol 7:265-276

Moldoveanu Z, Egan ML, Mestecky J (1984) Cellular origin of human polymeric and monomeric IgA: Intracellular and secreted forms of IgA. J Immunol 133:3156-3162

Nilsson K, Dahlgren UI, Hanson LÅ (1988) Origin of IgG and IgM antibodies appearing in milk and bile of rats (Manuscript in preparation)

Nilsson K, Dahlgren UIH, Hanson LÅ (1988) Different origin of IgA antibodies with various antigen specificities, appearing in rat bile (Submitted for publication)

Pierce PF, Kaper JB, Mekalanos JJ, Cray Jr WC, Richardson K (1987) Determinants of the immunogenicity of live virulent and mutant *Vibrio cholerae* in rabbit intestine. Infect Immun 55:477-481

Roberton DM, Carlsson B, Coffman K, Hahn-Zoric M, Jalil F, Jones C, Hanson LÅ (1988) Avidity of IgA antibody to *Escherichia coli* polysaccharide and diphtheria toxin in breastmilk from Swedish and Pakistani mothers (Submitted for publication)

Svennerholm A-M, Hanson LÅ, Holmgren J, Lindblad BS, Khan SR, Nilsson A, Svennerhom B (1981) Milk antibodies to live and killed polio vaccines in Pakistani and Swedish mothers. J Infect Dis 143:707-711

Svennerholm A-M, Hanson LÅ, Holmgren J, Lindblad BS, Nilsson B, Querrshi F (1980) Different secretory immunoglobulin A antibody responses to cholera vaccination in Swedish and Pakistani women. Infect Immun 30:427-430

Wold AE, Dahlgren UI, Hanson LÅ, Midtvedt T (1988) Antibody response in rats colonized with E. coli. I. Difference in response to bacterial and food antigens (Submitted for publication)

Wold AE, Dahlgren UIH, Ahlstedt S, Hanson LÅ (1987) Lack of IgA antibody response in secretions of rat dams during longterm ovalbumin feeding. Induction of systemic tolerance in pups but not in adult rats. Int Archs Allergy Appl Immunol 84:332-338

Systemic Immunization for the Induction of IgA Responses

A. Tarkowski[1,2], C. Lue[2], Z. Moldoveanu[2], H. Kiyono[2], J.R. McGhee[2], J.T. Prchal[3], N.B. Halpern[4], and J. Mestecky[2]

Departments of Microbiology[2], Medicine[3], and Surgery[4], University of Alabama at Birmingham, Birmingham, AL 35294, USA; [1] Present address: Department of Medical Microbiology, University of Göteborg, Sweden

INTRODUCTION

Although immune responses to bacterial polysaccharides are clinically important, the characterization of antibody responses from the point of view of induction, regulation, and isotypic patterns have not been carefully analyzed in humans. In contrast to protein antigens, the antibody response to polysaccharides has traditionally been regarded as thymus-independent (Howard et al. 1971; Fernandez and Möller 1977) although T cells may influence these responses (Baker et al. 1981). A major isotype of anti-polysaccharide antibodies is of the IgM isotype and is clonally restricted as shown by isoelectric focusing analysis (Insel et al. 1985). Proteins, on the other hand, are T cell-dependent antigens and induce immunologic memory with considerably less clonally restricted memory responses, and are largely restricted to the IgG class (Morrow et al. 1981).

Only limited information is available on the IgA response to polysaccharide antigens given by the systemic route. Kagnoff (1979) has established that in certain mouse strains IgA anti-dextran B1355 antibody production was age-dependent and of high titer, approaching those of the IgM isotype. Similar results have also been obtained by Eldridge and Meulbroek using type 3 pneumococcal polysaccharide (personal communication). Several groups (Kehrl and Fauci 1983; Heilman and Pedersen 1986; Munoz and Insel 1987; and Lue 1988) has reported that high frequencies of antigen-specific IgA-secreting cells occur after immunization with pneumococcal and *Haemophilus influenzae* type b polysaccharides, respectively, in humans.

The purpose of this study was to further characterize B cells committed to the production of anti-polysaccharide antibodies, to evaluate the compartmentalization of antigen-specific antibody production, to study molecular properties of the antibodies produced, and to assess the activation patterns of T cells during the immune response.

MATERIALS AND METHODS

Subjects and Vaccines

Healthy adult volunteers, 1 patient with a hepatobiliary stent and 1 patient with idiopathic thrombocytopenic purpura (ITP) were immunized with 0.5 ml of a 23 valent pneumococcal polysaccharide vaccine (Pnu-Imune, Lederle, Pearl River, N.Y., USA). Blood, saliva and tears were collected prior to vaccination and on 3-4 separate occasions between the 3rd and the 30th day after vaccination. Spleen and bone marrow specimens were obtained from the patient with ITP 6 days after vaccination. Bone marrow was obtained from one healthy individual by needle biopsy aspiration from the iliac crest 7

Current Topics in Microbiology and Immunology, Vol. 146

days after immunization. In addition, healthy individuals were immunized subcutaneously with tetanus toxoid (TT) (Wyeth Lab. Inc., Marietta, PA), meningococcal polysaccharides serotype A, C, Y, W-135 (Menomune, Connaught Lab. Inc., Swiftwater, PA), and with *Haemophilus influenzae* type b polysaccharide (Hib) (HIB-Imune, Lederle).

Cells and Body Fluids

Mononuclear cells were isolated by centrifugation on a Ficoll-Hypaque (Sigma, St. Louis, MO) density gradient, washed and resuspended in RPMI 1640 medium (GIBCO, Santa Clara, CA) supplemented with HEPES, sodium pyruvate, penicillin, streptomycin and 10% fetal calf serum (FCS) (complete medium). Bone marrow and spleen cells were treated with Tris-buffered 0.82% ammonium chloride solution at $0^{\circ}C$ in order to lyse erythrocytes, washed, and resuspended in complete medium. All mononuclear cells were processed in the ELISPOT assay or for fluorescence-activated cell sorter (FACS) analysis immediately after isolation. Unstimulated whole saliva, tears, bile, and serum were stored at -20° C until assayed.

Reagents

Affinity purified $F(ab')_2$ fragments of rabbit anti-human IgA (Pel-Freez, Rogers, AR), goat anti-human IgG (Jackson, West Grove, PA) and monoclonal mouse anti-human IgM (prepared in the hybridoma core facility of the Multipurpose Arthritis Center at UAB) were used as coating reagents in the ELISPOT and ELISA assays at concentrations of 5 µg/ml, 2 µg/ml and 15 µg/ml, respectively. Affinity-purified, biotinylated $F(ab')_2$ fragments of goat anti-human IgG, IgA, and IgM antibodies (Tago, Burlingame, CA) were used as developing antibodies in the ELISA and ELISPOT assays. Monoclonal mouse anti-human IgA1 and IgA2 (a generous gift from Dr. Jiri Radl, Rijswijk, The Netherlands) have been extensively tested for sensitivity and specificity in our laboratory (Russell et al. 1986). Biotinylated, affinity-purified goat-anti mouse IgG (Southern Biotechnology Associates, Birmingham, AL) extensively adsorbed with human immunoglobulins, was used as the second antibody when mouse monoclonal antibodies were used as primarily antibodies.

Assay for Enumeration of Antigen-Specific Antibody-Secreting Cells

The ELISPOT assay (Czerkinsky et al. 1983) was used to determine the frequencies of antibody-secreting cells (AbSC) to the polysaccharide and protein antigens used. Briefly, wells in the lids of 24 well tissue culture cluster plates (Costar, Cambridge, MA) were coated overnight at room temperature with polysaccharides previously coupled to poly-L-lysine according to a method described by Gray (1979). For the determination of anti-TT AbSC, the wells were coated with TT diluted in PBS. The plates were then washed with water and blocked with 5% FCS in PBS for 1 hour. One hundred microliters of mononuclear cell suspension was added to each coated well and incubated for 3 1/2 hr at 37° C. After another wash, biotinylated goat anti-human IgA, IgG, or IgM antibodies, diluted 1:750 in PBS containing 0.05% Tween 20 (PBS-Tween) were applied followed by incubation with 0.5 µg/ml of avidin-HRP (Sigma, St. Louis, MO) in PBS-Tween. For the detection of IgA subclasses 5 µg/ml of appropriate monoclonal antibody was used followed by stepwise incubation with biotinylated goat anti-mouse IgG (1:750 in PBS-Tween) and avidin-HRP. Following addition of the enzyme-substrate mixture (0.5 mg/ml of paraphenylenediamine and 0.015% H_2O_2 in 1% PBS-Noble agar (Difco,

Detroit, MI)) spots were enumerated under low magnification. Assays were always performed in duplicate and at several cell concentrations.

Assay for Determination of Immunoglobulin and Specific Antibody Levels in Serum, Bile, and External Secretions

The ELISA tests were performed in 96 well polystyrene microtiter plates (Dynatech, Alexandria, VA). The coating and developing reagents were used at the same concentrations as described above for the ELISPOT assay. However, the enzyme-substrate solution consisted of 2.5 mg/ml of 2,2 azinobis (3 Ethyl-benzthiazoline sulfonic acid) (Sigma, St. Louis, MO) in citrate buffer pH 4.2 containing 0.0075% H_2O_2. The absorbance was measured in a Titertek Multiscan photometer (Flow Laboratories, McLean, VA) at 414 nm.

Total IgA, IgG and IgM levels were quantitated by using a serum pool with known concentrations of isotypes as a standard. Purified IgA1 and IgA2 myeloma proteins were used as standards for the determination of total IgA subclass levels. The optical density (OD) values measured in the antigen-specific ELISA were compared to the OD values obtained from a reference pool of sera and salivas. All OD values were converted to antigen-specific ELISA units using calibration curves based on the OD obtained from serial dilutions of the reference serum and saliva, respectively. The calibration curves were constructed using a computer program based on either the 4-parameter logistic or weighted logit-log models (Russell et al. 1986). The calculations were carried out with the aid of a Macintosh SE computer (Apple Computer Inc., Cupertino, CA). The fold-increase (FI) values were calculated by dividing post-immunization ELISA unit values by the pre-immunization ELISA unit values (Pedersen and Henrichsen 1982; Lue et al. 1988).

Fluorescence-Activated Cell Sorting (FACS)

Peripheral blood mononuclear cells obtained on day 7 from one healthy volunteer immunized with pneumococcal polysaccharides were stained and sorted into surface IgA1 - positive ($sIgA1^+$), $sIgA2^+$, and transferrin receptor$^+$ (using monoclonal antibody produced by clone HB 21, American Type Culture Collection; ATCC) and $CD3^+$ (Leu 4, Becton-Dickinson, Mountain View, CA) cells and subsequently assayed in the antigen-specific ELISPOT assay.

Peripheral blood mononuclear cells from subjects immunized with pneumococcal polysaccharides, Hib, menigococcal polysaccharides and TT were analyzed with regard to the kinetics of major histocompatibility complex (MHC) class II antigen (HLA-DP, -DQ, -DR) expression on T cells.

RESULTS

Antibody Secretion by Peripheral Blood-, Spleen-, and Bone Marrow-Mononuclear Cells to Protein or Polysaccharide Antigens at the Single Cell Level

On the day of vaccination, no spontaneous anti-TT, anti-meningococcal, anti-Hib or anti-pneumococcal antibody-secreting cells could be detected in the peripheral blood of the subjects. On day 7, high frequencies of antigen-specific antibody-producing cells were observed for all of the various immunogens used. When protein immunogen was used, most cells secreted antigen-specific antibodies of the IgG class, whereas only few IgA- and no IgM-secreting cells could be detected. In contrast, when polysaccharide antigens were used, the overall pattern was a profound increase in antigen-specific antibody-secreting cells of the IgA isotype, whereas IgG and IgM increases were modest. A similar pattern, but of considerably less magnitude, was obtained when cells were analyzed for antigen-specific antibody production on days 4-5 and days 13-14, respectively, after immunization (Fig.1).

Analysis of antigen-specific IgA subclass expression at the single cell level indicated that a majority of IgA anti-protein antibodies were of the IgA1 isotype, whereas the majority of anti-polysaccharide antibodies were of the IgA2 subclass.

Anti-polysaccharide antibody-secreting cells were also detected in the bone marrow and spleen 6-7 days after vaccination. The isotypic pattern was similar in bone marrow and spleen when compared with PBMC responses.

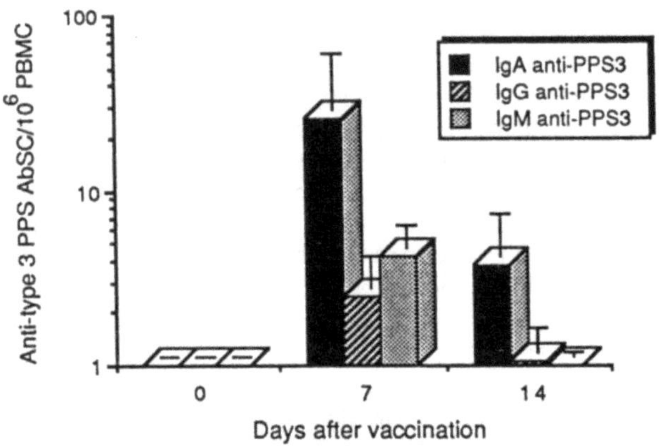

Fig. 1. Frequency of PBMC secreting antibodies to type 3 pneumococcal polysaccharide (PPS3). Results are expressed as geometric means x/+ SE of 10 individuals.

Increase of Antibody Levels in Serum and in External Secretions

Serum antibody levels increased significantly after immunization with both polysaccharide and protein vaccines. Peak responses were observed at day 14 and no significant change in antibody levels could be detected until day 30 after vaccination. IgA anti-polysaccharide antibodies displayed the highest increases while IgG and IgM anti-polysaccharide antibody responses were lower. The polymeric IgA2 fraction of anti-polysaccharide antibodies was responsible for most of the total IgA antibody response seen. Protein immunogens induced higher levels of IgG antibodies with lower responses seen in IgA or IgM isotypes. Total serum levels of IgG, IgA and IgM did not change significantly after immunization.

In saliva and tears the major increases of anti-polysaccharide antibodies were in the IgG class. Total levels of the IgG, IgA, and IgM in tears and saliva did not change significantly throughout the course of immunization.

Phenotypic Features of T- and B Cells Appearing in the Peripheral Blood after Polysaccharide Immunization

To study the phenotype of antigen-specific B cells, PBMC of one individual were double-stained with anti-transferrin receptor antibody and with anti-Leu 4 (CD3). The CD3 negative, transferrin receptor-positive cell fraction was enriched for antigen-specific antibody-secreting B cells. PBMC were also stained for surface IgA1 and IgA2, and separated by FACS. The fractions stained for IgA1 and IgA2 were enriched for IgA anti-polysaccharide antibody-secreting cells in percentages seen with the transferrin receptor-positive fraction.

The immunization regimen induced activation of T cells, as determined by the expression of MHC class II antigens, and was studied during the early stages of immunization. The frequency of MHC class II antigen-expressing T cells peaked 48 hours after immunization with the TT (protein antigen). Expression of MHC class II antigens on T cells developed slower after polysaccharide immunization and was of a considerably lower magnitude. In both polysaccharide and protein immunized subjects, the frequency of HLA-DP and HLA-DR-expressing T cells was higher when compared with the number of HLA-DQ positive T cells.

Hepatobiliary Clearance of Anti-Polysaccharide Antibodies

To examine the possible route of clearance of anti-polysaccharide antibodies, serum and bile were collected from one individual before and 14 days after pneumococcal immunization. Although the absolute levels of anti-pneumococcal antibodies were considerably lower in bile as compared to serum, the pattern in both compartments was strikingly similar.

DISCUSSION

This study has shown that systemic immunization with polysaccharide antigens induces a transient appearance of antibody-secreting cells in the peripheral blood of healthy adults. As shown in the present and previous studies (Kehrl and Fauci 1983; Heilmann and Pedersen 1986; Lue et al. 1988; Munoz and Insel 1987), peripheral blood B cells secreting IgA antibodies to pneumococcal polysaccharides or Hib account for the majority of antigen-specific antibody-producing cells. We obtained similar results using meningococcal polysaccharides of various serotypes. Taken together, the present results suggest that, in contrast to booster immunization with protein antigens which induces a large IgG antibody response (present study and Stevens and Saxon 1978; Stevens et al. 1979; Volkman et al. 1982; Tarkowski et al. 1986), polysaccharide vaccines result in responses of a different isotypic pattern. A predominant antigen-specific IgA-secreting cell response occurs following polysaccharide immunization and this was also observed in spleen and in bone marrow, suggesting a significant contribution by these two lymphoid organs for systemic immune responses. Strikingly high increases in the IgA anti-polysaccharide specific antibody levels in serum after immunization further support our results determined at the single cell level. These data favor the idea that a mucosal activation of polysaccharide-specific B cells occurs. Further support for this idea are the molecular properties of IgA anti-polysaccharide antibodies secreted after immunization. The majority of secreted IgA antibodies are J chain-containing polymers with a predominance of the IgA2 subclass, and are partially associated with secretory component (Lue et al. 1988); all three properties are characteristic for mucosally derived immunoglobulins.

The magnitude and isotype distribution of the antibody response in saliva differed from the response seen in serum. The increase rates of IgG and IgM anti-polysaccharide antibodies were quite similar to those seen in serum, whereas the increase of salivary antibodies of IgA class was lower when compared with serum increases. This indicates that, although polysaccharide immunization may activate IgA-committed mucosal B cells to secrete antibodies, as discussed above, these cells will not complete the migratory pathway by homing to distant mucosal sites, i.e., the salivary glands. Local IgG and IgM synthesis in the salivary glands could explain the increase in antibody levels observed in saliva after vaccination. Earlier studies indicated that systemic immunization of subjects previously exposed to antigen through the GI route results not only in systemic but also in secretory immune responses (Svennerholm et al. 1977). The relatively modest increases of secretory antibodies seen in our study could be due to different sites of the primary immunization (i.e., the respiratory or gastrointestinal tract) or, more likely, due to the different type of antigens used for the vaccination (protein vs polysaccharide).

It is well established that rodents eliminate IgA by hepatobiliary clearance (Jackson et al 1978; Delacroix et al. 1985). The present study indicates that hepatobiliary clearance of IgA takes place in humans as a physiological event after immunization with polysaccharides. The question arises as to whether IgA is cleared alone or in the form of immune complexes and should be further investigated. Nevertheless, this model system of studying hepatobiliary transport of IgA in humans is potentially of interest since IgA molecules studied are autologous, physiologically induced and unaltered by *in vitro* purification or other manipulations.

It has recently been shown that vaccination induced, antigen-committed PBMC are lymphoblastoid B cells and express transferrin receptor on their surface as an activation marker (Brieva et al. 1984). We confirmed these observations and showed that these transiently appearing cells still express surface Ig, according to their pre-plasma cell stage of differentiation (Lue et al. 1988). Of potential interest is the fact that *in vivo*- induced antigen-specific B cells can be highly enriched by sorting on FACS without affecting their functional capacity. This finding can be used to increase the efficacy of human-human hybridizations and/or EBV transformation procedures.

With regard to T lymphocytes, polysaccharide-specific immune responses are considered to be thymus-independent (Howard et al. 1971) although T cells are capable of modulating immune responses (Baker et al. 1981). Since IgA-mediated antibody responses are thought to be T-cell-dependent (Kawanishi et al. 1983; Kiyono et al. 1984) our finding that the majority of anti-PPS antibody producing cells secrete IgA is puzzling, indeed. Further studies are necessary to compare the *in vivo* activation properties of T cells after polysaccharide vaccination with the pattern after immunization with protein antigens.

The peripheral blood T cells were stained for MHC class II antigen expression. As previously reported (Yu et al. 1980), there was a peak increase in the frequency of class II antigen-positive T cells 48 h after vaccination with protein antigen. A delayed kinetic pattern and a lower magnitude of response were observed after polysaccharide vaccination. These observations suggest that T cells also participate *in vivo* in the polysaccharide responses. Due to the differences in kinetics and magnitude of MHC class II expression it may be suggested, however, that different T cell subsets, with possibly different functional capacities, are responsible for immune responses to protein and polysaccharide antigens, respectively.

REFERENCES

Baker PJ, Amsbaugh DF, Stashak PW, Caldes G, B. Prescott B (1981) Regulation of the antibody response to pneumoccal polysaccharide by thymus-derived cells. Rev Infect Dis 3:332-341

Brieva JA, Stevens RH (1984) Involvement of the transferrin receptor in the production and NK-induced suppression of human antibody synthesis. J Immunol 133: 1288-1292

Czerkinsky C, Nilsson L-Å, Nygren H, Ouchterlony Ö, Tarkowski A (1983) A solid-phase enzyme-linked immunospot (ELISPOT) assay for enumeration of specific antibody-secreting cells. J Immunol Methods 65:109-121

Delacroix DL, Malburny GN, Vaerman J-P (1985) Hepatobiliary transport of plasma IgA in the mouse: Contribution to clearance of intravascular IgA. Eur J Immunol 15:893-899

Fernandez C, Möller G (1977) Immune response against two epitopes on the same thymus-independent polysaccharide carrier. 1. Role of epitope density in carrier-dependent immunity and tolerance. Immunology 33:59-68

Gray B (1979) ELISA methodology for polysaccharide antigens: Protein coupling of polysaccharides for adsorption to plastic tubes. J Immunol Methods 28:187-192

Heilmann C, Pedersen FK (1986) Quantitation of blood lymphocytes secreting antibodies to pneumococcal polysaccharides after *in vivo* antigenic stimulation. Scand J Immunol 23:189-194

Howard JG, Christie GH, Courtnay BM, Leuchars E, Davies AJ (1971) Studies on immunological paralysis. VI. Thymic independence of tolerance and immunity to type III pneumococcal polysaccharide. Cell Immunol 2:614-626

Insel RA, Kittelberger A, Anderson P (1985) Isoelectric focusing of human antibody to the *Haemophilus influenzae* b capsular polysaccharide: Restricted and identical spectrotypes in adults. J Immunol 135:2810-2816

Kagnoff MF (1979) IgA anti-dextran B1355 responses. J Immunol 122:866-870

Kawanishi H, Saltzman LE, Strober W (1983) Mechanisms regulating IgA class-specific immunoglobulin production in murine gut-associated lymphoid tissues: I. T-cells derived from Peyer's patches that switch sIgM B cells to sIgA B cells in vitro. J Exp Med 157:433-450

Kehrl JH, Fauci AS (1983) Activation of human B lymphocytes after immunization with pneumococcal polysaccharides. J Clin Invest 71:1032-1040

Kiyono H, Cooper MD, Kearney JF, Mosteller LM, Michalek SM, Koopman WJ, McGhee JR (1984) Isotype specificity of helper T cell clones: Peyer's patch Th cells preferentially collaborate with mature IgA B cells for IgA responses. J Exp Med 159:798-811

Jackson CDF, Lemaitre-Coelho I, Vaerman JP, Bazin H, Beckers A (1978) Rapid disappearance from serum of intravenously injected rat myeloma IgA and its secretion into bile. Eur J Immunol 8:123-126

Lue C, Tarkowski A, Mestecky J (1988) Systemic immunization with pneumococcal polysaccharide vaccine induces a predominant IgA2 response of peripheral blood lymphocytes and increases of both serum and secretory anti-pneumoccal antibodies. J Immunol 140:3793-3800

Morrow CD, Macy EM, Stevens RH (1981) Analysis of antidiphtheria antibodies by isoelectric focusing: Evidence for restricted clonal heterogeneity of anti-fragment A antibodies. Infect Immun 31:1132-1137

Munoz JL, Insel RA (1987) *In vitro* human antibody production to the *Haemophilus influenzae* type b capsular polysaccharide. J Immunol 139:2026-2031

Pedersen FK, Henrichsen J (1982) Detection of antibodies to pneumococcal capsular polysaccharides by enzyme-linked immunosorbent assay. J Clin Microbiol 15:372-378

Russell MW, Brown TA, Radl J, Haaijman JJ, Mestecky J (1986) Assay of human IgA subclass antibodies in serum and secretions by means of monoclonal antibodies. J Immunol Methods 87:87-93

Stevens RH, Saxon A (1978) Immunoregulation in humans. Control of anti-tetanus toxoid antibody production after booster immunization. J Clin Invest 62:1154-1160

Stevens RH, Macy E, Morrow C, Saxon A (1979) Characterization of a circulating subpopulation of spontaneous antitetanus toxoid antibody producing B-cells following *in vivo* booster immunization. J Immunol 122: 2498-2504

Svennerholm A-M, Holmgren J, Hanson L-Å, Lindblad BS, Quereshi F, Rahimtoola RJ (1977) Boosting of secretory IgA antibody responses in man by parenteral cholera vaccination. Scand J Immunol 6: 1345-1349

Tarkowski A, Czerkinsky C, Nilsson L-Å (1985) Simultaneous induction of rheumatoid factor-and antigen-specific antibody-secreting cells during the secondary immune response in man. Clin Exp Immunol 61: 379-387

Volkman DJ, Allyn SP, Fauci AS (1982) Antigen-induced *in vitro* antibody production in humans. Tetanus toxoid-specific antibody synthesis. J Immunol 129:107-112

Yu TY, Winchester RJ, Fu SM, Gibofsky A, Ko HS, Kunkel HS (1980) Peripheral blood Ia-positive T cells. Increases in certain diseases and after immunization. J Exp Med 151:91-100

Prevention of Invasive Bacterial Diseases by Immunization with Polysaccharide-Protein Conjugates

J. B. Robbins, R. Schneerson, S. C. Szu, A. Fattom, Y. Yang, T. Lagergard, C. Chu, and U. S. Sørensen

Laboratory of Development and Molecular Immunity, National Institute of Child Health and Human Development, and the Pilot Production Plant, Laboratory of Cell and Molecular Biology, National Institute of Diabetes, Digestive and Kidney Diseases, National Institutes of Health, Bethesda, MD 20892, USA

INTRODUCTION

Invasive diseases, in otherwise healthy individuals, may be caused by both Gram-positive and Gram-negative bacteria which invade the bloodstream from their site on the mucous membranes of the human host. These bacteria include such diverse species as meningococci, pneumococcus, *Haemophilus influenzae* type b (Hib), and *Salmonella typhi* (typhoid bacillus). *Escherichia coli* K1 and Group B streptococci are major causes of disease in fetuses and in newborns. Their common characteristic is a capsular polysaccharide (CPS) that confers the property of invasiveness to these diverse bacteria. Unlike viral pathogens, each of these bacteria may induce several distinct diseases. For example, Hib may cause meningitis, epiglottitis, arthritis, pneumonia, etc. These varied clinical manifestations are, however, secondary to the bacteremia which antecedes tissue localization (Schneerson et al. 1987).

CAPSULAR POLYSACCHARIDES (CPS)

CPS are the most external surface structures of the bacteria in intimate contact with the host. In addition to their individual structure, CPS of pathogens have similar physico-chemical properties with each other and with the CPS of bacteria that are not pathogenic in healthy hosts:

1. CPS are polymeric structures composed of identical repeat units. Their complexity may be described as a linear homopolymer, linear copolymer, or linear branched chain copolymers;

2. CPS of invasive bacteria are composed of repeat units with one to six monosaccharides. Repeat units with more than five monosaccharides are usually not associated with pathogens (invasiveness is generally associated with CPS of comparatively simple composition);

3. Most CPS of invasive bacteria are negatively charged at neutral pH due either to a carboxylic acid or a phosphodiester bond in their repeat unit. The exceptions are pneumococcus type 7F and type 14 which do not have these moieties and are neutral;

4. CPS are "attached" to the bacterium by a different mechanism. Gram-negative CPS have a terminal glycolipid at their reducing end composed of a substituted glycerol with C1 bound to the aldehyde by a phosphodiester bond; the C2 and C3 positions

Current Topics in Microbiology and Immunology, Vol. 146

substituted by a fatty acid (Gotschlich et al. 1981). It is presumed that this hydrophobic moiety, at the reducing end of the CPS, inserts into the outer membrane similar to that of the Lipid A of LPS. The CPS of Gram-positive bacteria have not been as extensively investigated. There is evidence that the CPS of Group B streptococcus is covalently bound to the mucopeptide of the bacterium and the CPS of pneumococci is bound to the cell wall polysaccharide (C-polysaccharide) (Doran and Mattingly 1982; Sørensen et al. submitted for publication). The nature of these covalent bonds is not yet known;

5. Many CPS have moieties covalently bound via ester or ether linkages. These additional moieties include acetyl groups (N or O), pyruvate in O position, or amino acids linked via amino-hydroxyl. These moieties may affect the antigenic specificity and protective actions of the CPS as well as their susceptibility to enzymatic degradation (Schneerson et al. 1987);

6. CPS have a less "ordered" structure than globular proteins and are mixtures of molecules with varying length. Accordingly, it has been difficult to describe their molecular weight. The immunogenicity of CPS is related to their molecular weight or "size" (Kabat and Bezer 1958; Wong et al. 1977). Standardization of CPS vaccines has relied upon their gel filtration profiles to assure vaccines of a minimum "molecular size" and, therefore, predictable immunogenicity.

Many cases illustrate that the structure is related to the virulence-promoting and immunologic properties of CPS. For example, only type b of the six *H. influenzae* CPS are associated with invasive diseases (Pittman 1931). Similarly, only 5 of the 14 meningococcus CPS and about 23 of the 86 pneumococcus CPS are associated with invasive diseases caused by these bacteria. It is not yet possible, however, to predict which of these CPS will be associated with invasiveness on the basis of their structure. The best explanation for their virulence properties is that only some of the many CPS serve as effective "shields" against the protective actions of complement. Non-capsulated Gram-negative bacteria are readily lysed *in vitro* by complement alone; their LPS serves to activate the C1q component (Schreiber et al. 1979). Complement proteins up to C3 or C5 are deposited in active form upon non-capsulated Gram-positive bacteria in the absence of antibody (Brown et al. 1983). The ability of complement alone to activate lysis, opsonization, etc., explains the "natural" resistance of humans to most bacteria: only certain CPS form "effective shields" against the protective actions of complement (Sutton et al. 1982). It follows that the CPS of non-invasive bacteria are ineffective "shields". For non-capsulated invasive bacteria, such as *S. typhimurium*, the non-reducing end of the 0-specific side chain of the LPS exerts this "shielding" effect (Leive and Limenez-Lucho 1987). Protective immunity against invasive diseases caused by capsulated bacteria, therefore, requires serum antibodies capable of inducing protective complement actions. The relation between the structure and the shielding ability of CPS remains a tantalizing question.

MUCOSAL INTERACTION WITH INVASIVE CAPSULATED BACTERIA

Three aspects concerning this phase of host-capsulated pathogen interaction will be discussed. First, there is a lack of quantitative data about colonization of the mucous membranes by capsulated pathogens. Reports cite only the detection of pneumococci, meningococci, or *H. influenzae* type b, in the nasopharynx and in the stool of *S. typhi* and *E. coli*, or of Group B streptococci in the vagina, cervix, and uterine tissues. It has not been possible, therefore, to compare the outcome of colonization, i.e., bacteremia, with quantitation of capsulated bacteria at these various sites. Further, no attempts have been made to distinguish between individuals who may be harboring the organisms without apparant disease from those who are "spreaders". The use of selective and differential media, such as may be achieved with antibiotics and typing antiserum, could provide a powerful tool for examining this problem (Michaels et al. 1975).

The second point concerns the mechanism by which capsulated bacteria interact with mucous membranes. Pili have been demonstrated on Gram-negative capsulated pathogens including meningococci, Hib, *E. coli* K1 and *S. typhi*. The pili of *E. coli* consist of several allelic genes and it is not yet clear which of the several families mediates adherence of the organism to the mucous membranes. Further, it is difficult to demonstrate pili on strains of capsulated bacteria isolated from the blood or cerebrospinal fluid (Craven et al. 1980). This latter finding indicated that pili may undergo phenotypic changes and that there is selective pressure for variants to be expressed in various tissues.

The last point is that colonization of the mucous membranes with these capsulated pathogens is almost always asymptomatic. Meningococci, Hib, and pneumococci do not induce inflammation of the nasopharynx. Similarly, neither invasive *E. coli* K1 strains or *S. typhi* cause inflammation of the small intestine's mucous membranes. Infection and rupture of the Peyer's patches occurs through secondary infection of these tissues by *S. typhi* strains from the blood (Robbins and Robbins 1984). The mechanism by which Gram-positive capsulated bacteria establish colonization on respiratory mucous membranes is unclear. Vaccine-induced CPS antibodies, however, have been shown to inhibit acquisition of colonization with pneumococci and meningococci (MacLeod et al. 1945; Goldschneider et al. 1973). Passive immunization with CPS antibodies did not eliminate established colonization of the mucous membranes with capsulated pathogens. This suggests that serum CPS antibodies exert only a minor (if any) role, in altering the interaction of capsulated pathogens with mucous membranes.

Two immunologic properties of CPS injected into humans limit their universal use as vaccines. First, is their lack of T cell-dependent immunogenicity. With the exception of Group A meningococcal CPS in infants, reinjection of CPS does not induce a booster antibody response (there is no recruitment of T-helper cells by the CPS-stimulated B cells). Second, is their age-related immunogenicity. Most invasive diseases caused by capsulated bacteria have their highest incidence in infants and young children: CPS of these bacteria fail to induce a protective serum antibody in this age group (Robbins et al. 1983). Further complicating this problem is the observation that the duration of protective antibody levels induced by CPS is shortest in the young: reinjection only serves to induce a primary response characteristic of the host's age (Heidelberger et al. 1950; Gold and Lepow 1976; Kayhty et al. 1984). Third, patients with chronic diseases, especially malaria, and malnutrition, respond less to CPS vaccines (Greenwood 1984). Further, the duration of vaccine-induced antibodies and protection in this latter group of patients is shorter than in healthy individuals (Reingold et al. 1985).

CAPSULAR POLYSACCHAIRDE-PROTEIN CONJUGATES

Most activity has been directed towards studying Hib conjugates (Schneerson et al. 1987). Hib, a Gram-negative bacterium, is pathogenic for humans only: a rarely encountered organism is most often found in the upper respiratory tract of infants and children. In non-immune individuals, Hib may invade the blood and multiply. If its blood stream concentration $\geq 10^3$ organisms/ml, Hib may penetrate and infect the meninges or other tissues (Moxon and Murphy 1978). In the United States, it has been proposed that about 1 in 250 infants and children contract meningitis due to this organism each year (Parke et al. 1972). The incidence is higher in selected populations such as native Americans and in patients with splenic dysfunction such as sickle cell anemia, asplenia, and hereditary spherocytosis. Despite effective antibiotics and supportive therapy, 5 to 10% of patients succumb and 30 to 50% of those who recover endure permanent central nervous system damage (Schneerson 1988). Epiglottitis, a serious and potentially fatal upper respiratory infection, is also caused by Hib (Claesson et al. 1984).

Tetanus toxoid (TT) was chosen for the carrier protein of Hib conjugates because of its history as an effective vaccine. TT was passed through S-300 Sephacryl to remove the high molecular weight aggregates and lower molecular weight degradation products. Hib CPS was reacted with cyanogen bromide to form the isocyanate intermediate which readily binds to the bifunctional nucleophile, adipic acid dihydrazide. The adipic hydrazide derivative of Hib CPS and TT are covalently bound by carbodiimide-mediated condensation and further purified by gel filtration. This scheme has been used to synthesize conjugates with pneumococcus types 6A and 6B and *E. coli* K13 CPS (Schneerson et al. 1980, 1984, 1986, 1987; Chu et al. 1983).

Hib-TT has been evaluated in both inbred and outbred laboratory mice and in juvenile and infant rhesus monkeys injected alone or concurrently with pneumococcus type 6A-TT, or *E. coli* K100-TT, or TT alone or formulated in DTP. The results are summarized:

1. Hib CPS was non-immunogenic in most mice and in juvenile and infant rhesus monkeys (other CPS are more immunogenic in humans than in other species). Hib-TT, in contrast, elicited protective levels of Hib CPS antibodies in mice and in juvenile and infant rhesus monkeys. Antibody responses were dose-dependent, increased with carrier priming or with incorporation into complete Freund's adjuvant, and could be boosted with second and third injections. Hib CPS, therefore, has both increased immunogenicity and T-dependent properties as a conjugate with TT (Claesson et al. 1984; Schneerson et al. 1980, 1984, 1986; Chu et al. 1983);

2. Concurrent injection of Hib-TT and the cross-reacting pneumococcus type 6A-TT enhanced the Hib antibody response. No enhancement or depression was observed on the type 6A antibody response in adult humans. It follows that two conjugates can be injected in one formulation;

3. Concurrent injection of TT along with Hib-TT accelerated the Hib antibody responses in both adult and infant rhesus monkeys ("carrier" effect). Hib-TT and Pn6A-TT could be incorporated into DTP formulation as part of routine infant immunization (Table 1);

4. Passive immunization of juvenile rhesus monkeys with an adult human dose of tetanus immune globulin one day prior to immunization with Hib-TT did not affect the Hib antibody responses (Schneerson et al. 1984). Siber et al. (1984) reported that maternally-derived Hib antibodies did not reduce the antibody response elicited by another Hib conjugate. Maternally-derived antibodies were shown to exert slight, but statistically significant epitope-specific suppression in 3-months old infants injected with Hib-TT (Claesson et al. 1988). Differences in the secondary biological properties of

Cohn fraction II immunoglobulin and maternally-derived antibodies could explain this difference.

Table 1. *Haemophilus influenzae* (Hib) antibodies elicited (µg Ab/ml) in juvenile rhesus elicited by Hib conjugates alone or with tetanus toxoid (TT), DTP, tetanus immune globulin (TIG) or with cholera toxin (CT) as the carrier protein, [n=8]

Vaccine	Pre-immun	1st inj	2nd inj	3rd inj	7 wk TT	2 wk post TT
Hib-TT	0.06	1.22	2.15	2.27	2.18	1.80
Hib-TT + TT	0.10	2.18	4.38	7.29	7.32	5.40
Hib-TT + DTP	0.05	4.65	14.61	20.65	13.04	7.25
Hib-TT + TIG	0.08	2.00	2.45	3.78	2.78	1.95
Hib-CT	0.06	2.75	6.91	10.35	5.57	4.25
Hib	0.06	0.04	0.05	<0.04	0.05	0.04
TT	0.05	0.05	0.06	0.06	0.06	0.06

Following successful experiments with laboratory animals, Hib-TT was clinically evaluated alone, or in combination with pneumococcus type 6A or *E. coli* K100 CPS in young adult volunteers (Schneerson et al. 1986). Local reactions were common and were probably due to Arthus reactivity mediated by pre-existing TT antibodies. Fever occurred in about 10% of the volunteers after the first injection only. Hib-TT elicited about a 180-fold increase in Hib antibodies and the Pn6A-TT conjugate elicited about a 8-fold increase in Pn6A CPS antibodies after 1 injection (Table 2). Booster reactions were not elicited in adults. A one-way cross-reaction was noted as Pn6A conjugates elicited 2-fold or greater increases in the levels of Hib CPS antibodies in 13/20 of the volunteers; only 4/59 immunized with Hib-TT had an increase in Pn6A antibodies. Pre-immunization Hib antibodies were composed of IgM, IgA, and IgG immunoglobulins. Post-immunization sera showed increases in all three isotypes; elevation of IgG was the highest of the three isotypes. Conjugate-induced antibodies to both the CPS and TT exerted secondary biological activities that have been correlated with immunity. All 73 volunteers examined maintained a 4-fold or greater elevation of Hib. All volunteers maintained a $\geq$ 4-fold increase of Hib antibodies six months after their second immunization. The post-immunization levels declined in six months to about 80% of the maximum attained by immunization. The fold increases and post-immunization geometric mean antibody levels elicited by the conjugates, used in our experiments and those synthesized by other methods, were about five times higher than those elicited by the Hib CPS confirming the increased immunogenicity of the Hib-TT. The lack of a booster response in the adults was likely due to the high levels of antibodies to both the protein and CPS components elicited by the conjugates (probably the maximal attainable by subcutaneously injected saline solutions). Both Hib, CPS, and Hib conjugates induced an increase in IgM, IgA, and IgG isotypes; IgG was the predominate isotype induced by both. The level of IgG Hib antibodies declined the most after six months. The IgG subclasses and the "clonotype", as revealed by isoelectric focusing, are similar in recipients of either the Hib CPS or the conjugate (Moxon and Murphy 1978). The relative homogeneity of CPS antibodies compared to the heterogeneity of anti-protein antibodies, reported by Yount et al. (1968), is observed in antibodies elicited by CPS as well as by CPS-protein conjugates: the difference between the serum antibodies elicited by the two immunogens seems to be quantitative, and not qualitative. Accordingly, it appears that both the Hib-TT are stimulating lymphoid cells previously differentiated to synthesize antibodies of a single specificity. Anderson et al. (1986) prepared conjugates with periodate-oxidized products of Hib CPS. These two conjugates exhibited increased immunogenicity and T cell-dependence in adults and children: only the conjugate prepared with CPS of 20 or greater repeating units induced protective levels of Hib CPS antibodies consistently in infants. Antibodies to both Hib CPS and diphtheria toxin elicited by this conjugate had secondary biological properties correlated with immunity. A detailed analysis of Hib antibodies elicited in infants by the conjugates has been reported by these workers (Schneerson et al. 1986). Using the technique of isoelectric focusing and radioautography with ^{125}I-labeled Hib CPS, the authors have demonstrated that serum Hib CPS antibodies elicited by the conjugate or the CPS alone had similar oligoclonal properties.

Table 2. Serum *Haemophilus influenzae* type b capsular polysaccharide antibodies (μg Hib Ab/ml) in young adult volunteers injected with conjugates composed of Hib, pneumococcus type 6A (Pn6A) bound to tetanus toxoid (TT) alone or concurrently

	Geometric mean and 25th and 75th percentiles			
Group	Pre inject	3 wks post 1st inject	2 wks post 2nd inject	6 mos post 1st inject
50 μg Hib-TT injected two times	1.12 (0.55-2.53)	202.2 (95.8-422)	162.0 (71.5-350)	84.2 (48.6-166)
50 μg Pn6A-TT injected two times	1.61 (0.59-3.38)	4.3 (1.81-10.2)	5.90 (2.14-11.4)	5.04 1.57-11.3)
1st; 50 μg Hib-TT and Pn6A-TT, 2nd; 50 μg Hib-TT	0.98 (0.32-1.66)	147.7 (96.5-241)	152.3 (97.0-253)	89.4 (56.0-140)
1st; 100 μg Hib-TT, 2nd; 50 μg Hib-TT	1.18 (0.33-4.52)	235.7 (149-410)	206.8 (131-430)	83.6 (49.6-166)

In their original experiments, Avery and Goebel (1929), reported Pn type 3 CPS formed a more immunogenic conjugate than the disaccharide repeating unit. An optimal immunogenic size of the PS component of the conjugates was suggested by Makela et al. (1984). Immunization of mice with conjugates prepared with dextrans of high, intermediate and low molecular weights showed that the highest levels of anti-dextran antibodies were elicited by dextran of intermediate size (circa 40,000). Interpretation of these data must consider the difficulties in preparing conjugates with dextrans of varying size as their only variable. The extent of derivatization of the dextran with adipic acid dihydrazide, the ratio of protein to carbohydrate, and the molecular weights of these conjugates, prepared with dextrans of varying size, were different. We have not explored this approach extensively. Our preliminary results show that conjugates composed of CPS of higher molecular weight are better immunogens; a decrease in immunogenicity with our highest molecular weight Hib CPS, Pn6A CPS and the Vi CPS of *S. typhi* was not observed (Schneerson et al. 1987).

CARRIER PROTEINS

Information gained by the study of haptens is applicable to understanding the immunological properties of conjugates. Covalent binding to a protein imparts new immunological properties to both haptens and CPS (Schneerson et al. 1987; Claesson et al. 1988). The immunogenicity of the carrier protein is an important factor in determining the amount of antibody directed towards the hapten or PS. For example, little or no anti-hapten antibody is elicited by conjugates prepared with homologous serum proteins. The results obtained with conjugates prepared with horseshoe crab haemocyanin (HCH) and Hib CPS provide further evidence. HCH elicited high levels of specific antibodies in mice and was a good carrier protein for conjugates with Hib (Schneerson et al. 1980). Unexpectedly, juvenile rhesus monkeys responded poorly to this protein antigen and conjugates of HCH and the Hib CPS elicited comparatively low levels of antibodies (Chu et al. 1983). No increase in the immunogenicity of Hib CPS was observed when it was bound to Pn type 3 CPS (Claesson et al. 1984), another T cell-independent antigen. A T cell-dependent response to dinitrophenyl was not elicited by this hapten conjugated to Pn type 3 CPS (Mitchell et al. 1972). The immune response to the PS (hapten) elicited by conjugates still retains the restricted nature of the anti-PS antibodies elicited when CPS is used alone and is preserved even when induced by PS-protein conjugates. Antibodies elicited by conjugates are qualitatively similar but quantitatively higher than those elicited by the CPS (Schneerson et al. 1987; Claesson et al. 1988). In our clinical studies, pre- and post-immunization levels of TT and Hib CPS antibodies were unrelated (Schneerson et al. 1986).

PROTECTIVE EFFECT OF *H. influenzae* TYPE B CONJUGATES

Recently, Eskola et al. (1987) reported that immunization of infants with 4 injections of Hib CPS-diphtheria toxoid (DT) conferred about 80% protection in children 11 months of age and older, against systemic diseases caused by this organism. This Hib-DT conjugate was prepared according to the original scheme of Schneerson et al. (1980). Currently, an effectiveness trial of this same conjugate in Alaskan Eskimo infants is ongoing (the attack rate is considerably higher and age-distribution of Hib meningitis in this population is lower than in most infants and children in the USA, with about 50% of the cases occurring in infants six months (or less) of age (Ward et al. 1986).

Vi CONJUGATES FOR THE PREVENTION OF TYPHOID FEVER

Typhoid fever remains an important cause of morbidity and mortality in countries where effective control of sewage disposal, distribution of food, and identification of patients has not yet been achieved. Development of vaccines for prevention of typhoid fever has been difficult since the causative agent, *S. typhi*, is a pathogen for and a habitant of humans only - there is no suitable animal model. Information gained from animal studies, therefore, may have only limited application to the understanding of typhoid fever (Hornick et al. 1970).

The pathogenesis and immunity to typhoid fever is complex: both constitutive and acquired resistance mechanisms of the mucosal and systemic tissues are involved. Most cases occur following ingestion of drinking water, or occasionally food, contaminated by the *S. typhi*. Organisms that survive the gastric acidity establish asymptomatic colonization of the small intestine. In non-immune individuals, *S.*

typhi may penetrate the intestinal mucosa into the blood stream. Those organisms that survive start to grow and cause septicemia with its fever, shock-like states, and leukopenia: these symptoms are probably mediated by the LPS of the pathogen. It is of interest that the LPS of *S. typhi* has very high specific "endotoxic activities" when compared to those of other Gram-negative pathogens (Greisman and Hornick 1969). The complications of typhoid fever, including infection and rupture of the Peyer's patches with perforation, are secondary to the bacteremia.

While sharing most of the genome of the species, *S. typhi* has several metabolic traits and a capsular polysaccharide, the Vi antigen, which distinguish it from other *Salmonellae* (Farmer et al. 1985). We feel it is the Vi which confers higher virulence (higher incidence, morbidity, and mortality) to *S. typhi* compared to the enteric fevers caused by the other *Salmonellae*. The Vi probably acts as a "shield", protecting the pathogen from the actions of serum complement, similar to the CPS of other capsulated bacterial pathogens. Serum Vi antibodies confer immunity by facilitating formation of protective serum complement sites on *S. typhi* (Robbins et al. 1988). This does not imply that serum Vi antibodies are the only acquired protective mechanism. The use of the attenuated *S. typhi* strain, Ty21A, has suggested that secretory anti-**O** antibodies of the IgA immunoglobulin class may facilitate opsonization of *S. typhi* by monocytes within the intestinal tract (Tagliabue et al. 1986).

Previous studies by Landy (1954), provided evidence for both the virulence-promoting and protective activity of the Vi in animal models. When used as an experimental vaccine, however, the Vi did not elicit protective immunity in volunteers challenged with *S. typhi* (Hornick et al. 1970). These workers were obliged to treat their Vi with 1N acetic acid at $100^{\circ}C$, for 24 hours to remove the "endotoxic" activity of their preparation. Technology was not available to appreciate the extensive denaturation to both the immunogenicity (depolymerization) and antigenicity (removal of the **O**- and **N**-acetyl moieties) that resulted from this treatment (Robbins and Robbins 1984).

Two, randomized, double blinded, clinical trials in areas with high attack rates of typhoid (about 1%) showed that Vi prepared under non-denaturing conditions conferred about 70% immunity against typhoid for at least 2 years (Acharya et al. 1987; Klugman et al. 1987). Volunteers in these two areas showed a lesser response (measured by $\geq$ 4-fold increase) to the Vi than young adults in France and the U.S. This finding prompted us to synthesize conjugates with this protective antigen (Szu et al. 1987). A synthetic scheme, using the carboyxl function to attach a sulphydryl group (-SH) to the Vi, was devised. SH-Vi was bound to several proteins, including TT and the beta subunit of cholera toxin. These Vi conjugates were shown to be more immunogenic and to elicit booster responses in mice and in rhesus monkeys (Table 3). Clinical studies of these new vaccines are planned.

SUMMARY

Covalent binding of CPS to T cell-dependent carrier proteins to form conjugates can be done by clinically acceptable methods. As a component of a conjugate, two immunologic properties of CPS are changed: 1) their immunogenicity is increased and; 2) reinjection induces a booster response in the young (T cell-dependence). Serum antibodies induced by the CPS alone, or as a component of a conjugate, are qualitatively similar: the difference between antibodies elicited by the CPS or the conjugate is quantitative. A clinical trial with a Hib-DT conjugate showed that conjugates could confer immunity in an age group not protected by the CPS alone.

Table 3. Serum antibodies elicited in juvenile Rhesus monkeys with Vi, Vi-cholera toxin or pneumococcus type 6B-tetanus toxoid conjugates

		µg Ab/ml geometric mean (80% confidence limits)		
Vaccine	n=	Pre-immune	1st injection	2nd injection
Vi	6	0.07 (0.05-0.10)	0.23[d] (0.06-0.82)	0.07[e] (0.05-0.09)
Vi-CT_{XII}	8	0.07[a] (0.03-0.19)	0.93[b] (0.24-3.55)	3.65[c] (0.72-18.5)
Pn6B-TT	8	0.08 (0.06-0.11)	0.10 (0.05-0.21)	0.10 (0.04-0.23)

a vs b: p=0.004, b vs c: p= 0.02, a vs c: p=0.0003,
b vs d: p=0.009, c vs e: 0.001.

Induction of serum CPS antibodies confers protection against capsulated bacteria in the bloodstream: their role in the interaction of these pathogens on the mucous membranes has not been characterized. Preliminary *in vitro* experiments suggest that secretory antibodies to non-capsular structures may also exert protective immunity.

REFERENCES

Acharya IL, Lowe CU, Thapa R, Gurubacharya VL, Shrestha MB, Cadoz M, Schulz D, Armand J, Bryla DA, Trollfors B, Cramton T, Schneerson R, Robbins JB (1987) Prevention of typhoid fever in Nepal with the capsular polysaccharide of *Salmonella typhi*. N Engl J Med 317:1101-1104

Anderson PW, Pichichero ME, Insel RA, Betts R, Eby R, Smith DH (1986) Vaccines consisting of periodate-cleaved oligosaccharides from the capsule of *Haemophilus influenzae* type b coupled to a protein carrier: Structural and temporal requirements for priming in the human infant. J Immunol 137:1181-1186

Avery OT, Goebel WF (1929) Chemo-immunological studies on conjugated carbohydrate-proteins. II. Immunological specificity of synthetic sugar-proteins. J Exp Med 50:521-533

Brown EJ, Joiner KA, Gaither TA, Hammer CH, Frank MM (1983) The interaction of C3b bound to pneumococci with factor H (b1H globulin), factor 1 (C3b/C4b inactivator) and properdin factor B of the human complement system. J Immunol 131:409-415

Chu CY, Schneerson R, Robbins JB, Rastogi SC (1983) Further studies on the immunogenicity of *Haemophilus influenzae* type b and pneumococcal type 6A polysaccharide-protein conjugates. Infect Immun 40:245-256

Claesson B, Trollfors B, Ekstrom-Jodal B, Jeppsson P-H, Lagergard T, Nylen O, Rigner P (1984) Incidence and prognosis of acute epiglottitis in children in a Swedish region. Pediat Infect dis 3:534-538

Claesson BA, Trollfors B, Lagergard T, Cramton T, Bryla D, Otterman G, Schneerson R, Robbins JB (1988) Clinical and immunological responses to the capsular polysaccharide of *Haemophilus influenzae* type b alone or conjugated to tetanus toxoid in 18- to 23-month-old children. J Pediatr 112:695-702

Claesson BA, Schneerson R, Robbins JB, Johansson J, Lagergard T, Taranger J, Bryla D, Levi L, Cramton T, Trollfors B (in press) Protective levels of serum antibodies stimulated in 3 month old infants by two injections of *Haemophilus influenzae* type b capsular polysaccharide tetanus toxoid (Hib-TT) conjugates. J Pediatrics

Craven DE, Frasch CE, Peppler MS, Mocca LF, McGrath PP, Washington G (1980) Adherences of *Neisseria meningitidis* from patients and carriers to buccal epithelial cells. J Infect Dis 142:556-568

Doran TI, Mattingly SJ (1982) Association of type- and group-specific antigens with the cell wall of serotype III Group B streptococcus. Infect Immun 36:1115-1122

Eskola J, Peltola H, Takal AH, Kayhty H, Hakulinen M, Karanko V, Kela E, Rekola P, Ronnberg PR, Sameuelson JS, Gordon LK, Makela PH (1987) Efficacy of *Haemophilus influenzae* type b polysaccharide-diphtheria toxoid conjugate vaccine in infancy. N Engl J Med 317:717-721

Farmer JJ, Davis BR, Hickman-Brenner FW, McWhorter A, Huntley-Carter GP, Asbury MA, Riddle C, Wathen-Grady HG, Elias E, Fanning GR, Steigerwalt AG, O'Hara CM, Morris GK, Smith PB, Brenner DJ (1985) Biochemical identification of new species and biogroups of *Enterobacteriaceae* isolated from clinical specimens. J Clin Microbiol 21:46-76

Gold R, Lepow ML (1976) Present status of polysaccharide vaccines in the prevention of meningococcal disease. Adv Pediatr 23:71-93

Goldschneider I, Lepow ML, Gotschlich EC, Mauck FT, Bachl F, Randolph M (1973) Immunogenicity of Group A and Group C meningococcal polysaccharides in human infants. J Infect Dis 128:769-776

Gotschlich EC, Fraser BA, Nishimura O, Robbins JB, Liu T-Y (1981) Lipid on capsular polysaccharides of gram-negative bacteria. J Biol Chem 256:8915-8921

Greenwood BM (1984) Selective primary health care: Strategies for control of disease in the developing world. XIII. Acute Bacterial Meningitis. Rev Infect Dis 6:S374-S389

Greisman SE, Hornick RB (1969) Comparative pyrogenic reactivity of rabbit and man to bactrial endotoxin. Proc Soc Exp Biol Med 131:1154-1158

Heidelberger M, DiLapi MM, Siegel M, Walter AW (1950) Persistance of antibodies in human subjects injected with pneumococcal polysaccharides. J Immunol 65:535-541

Hornick RB, Greisman SE, Woodward TE, DuPont HL, Dawkins AT, Snyder MJ (1970) Typhoid fever: Pathogenesis and immunologic control (second of two parts) N Engl J Med 283:739-746

Kabat EA, Bezer AE (1958) The effect of variation in molecular weight on the antigenicity of dextran in man. Arch Biochem Biophys 78:306-310

Kayhty H, Karanko V, Peltola H, Makela PH (1984) Serum antibodies after vaccination with *Haemophilus influenzae* type b capsular polysaccharide. No evidence for tolerance or memory. Pediatrics 74:857-865

Klugman KP, Koornhof HJ, Gilbertson IT, Robbins JB, Schneerson R, Schulz D, Cadoz M, Armand J, Vaccine Advisory Committee (1987) Protective activity of Vi capsular polysaccharide vaccine against typhoid fever. Lancet ii:1165-1169

Landy M (1954) Studies in Vi antigen. (VI. immunization of human beings with purified Vi antigen.) Amer J Hyg 60:52-62

Leive L, Limenez-Lucho VE (1987) Lipopolysaccharide O-antigen structure controls alternative pathway activation of complement: Effects on phagocytosis and virulence of *Salmonellae*. American Society for Microbiology, Abstract, p 14

MacLeod CM, Hodges RG, Heidelberger M, Bernhard WG (1945) Prevention of pneumococcal pneumonia by immunization with specific capsular polysaccharides. J Exp Med 82:445-465

Makela O, Peterfly F, Outschoorn JG, Richter AW, Seppala I (1984) Immunogenic properties of alpha (1-6) dextran, its protein conjugates and conjugates of its breakdown products in mice. Scand J Immunol 19:541-550

Michaels RH, Stonebreaker FE, Robbins JB (1975) Use of antiserum agar for detection of *Haemophilus influenzae* type b in the pharynx. Pediatr Res 9:513-516

Mitchell GF, Humphrey JH, Williamson AR (1972) Inhibition of secondary anti-hapten responses with the hapten coupled to type 3 pneumococcal polysaccharide. Eur J Immunol 2:460-464

Moxon R, Murphy PA (1978) *Haemophilus influenzae* bacteremia and meningitis resulting from the survival of a single organism. Proc Natl Acad Sci USA 75:1534-1536

Parke JC, Schneerson R, Robbins JB (1972) The attack rate, age incidence, racial distribution and case fatality rate of *Haemophilus influenzae* type b meningitis in Mecklenburg County, North Carolina. J Pediat 81:765-769

Pittman M (1931) Variation and type specificity in the bacterial species *Haemophilus influenzae*. J Exp Med 53:471-492

Reingold AL, Broome CV, Hightower AW, Ajello GW, Bolan GA, Adamsbaum C, Jones EE, Philips C, Tiendrebeogo H, Yada A (1985) Age-specific differences in duration of clinical protection after vaccination with meningococcal polysaccharide A vaccination. Lancet ii:114-118

Robbins JB, Austrian R, Lee C-J, Rastogi SC, Schiffman G, Henrichsen J, Makela PH, Broome CV, Facklam RR, Tiesjema RH, Parke Jr JC (1983) Consdiderations for formulating the second-generation pneumococcal capsular polysaccharide vaccine with emphasis on the cross-reactive types within groups. J Infect Dis 148:1136-1159

Robbins JD, Robbins JB (1984) Re-examination of the immunopathogenic role of the capsular polysaccharide (Vi antigen) of *Salmonella typhi*. J Infect Dis 47:436-499

Robbins JB, Schneerson R, Acharya IL, Lowe CU, Szu SC, DAniels E, Yang Y, Trollfors B (1988) Protective roles of mucosal and serum immunity against typhoid fever. Monogr Allerg 24:315-320

Schneerson R, Barrera O, Sutton A, Robbins JB (1980) Preparation, characterization and immunogenicity of *Haemophilius influenzae* type b polysaccharide-protein conjugates. J Exp Med 152:361-376

Schneerson R, Robbins JB, Chu C-Y, Sutton A, Vann W, Vickers JC, London WT, Curfman B, Hardegree MC, Shiloach J, Rastogi SC (1984) Serum antibody responses of juvenile and infant rhesus monkeys injected with *Haemophilus influenzae* type b and pneumococcus type 6A polysaccharide-protein conjugates. Infect Immun 45:582-591

Schneerson R, Robbins JB, Parke Jr JC, Bell C, Schlesselman JJ, Sutton A, Wang Z, Schiffman G, Karpas A, Shiloach J (1986) Quantitative and qualitative analyses of serum antibodies elicited in adults by *Haemophilus influenzae* type b and pneumococcus type 6A capsular polysaccharide-tetanus toxoid conjugates. Infect Immun 52:519-528

Schneerson R, Robbins JB, Szu SC, Yang Y (1987) Vaccines composed of polysaccharide-protein conjugates: Current status, unanswered questions, and prospects for the future. In: R Bell, G Torrigiani (eds) Towards Better Carbohydrate Vaccines. World Health Organizaton. John Wiley & Sons, Ltd. United Kingdom, p 307

Schneerson R (1988) Vaccines for prevention of mental retardation following meningitis. In: JF Kavanagh (ed) Understanding Mental Retardation. Paul H. Brookes, Baltimore, p 237

Schreiber RD, Morrison DC, Podack ER, Müller-Eberhard HJ (1979) Bactericidal activity of the alternative component pathway generated from 11 isolated plasma proteins. J Exp Med 149:870-882

Siber G, Ambrosino DM, McIver J, Ervin TJ, Schiffman G, Sallan S, Grady GF (1984) Preparation of human hyperimmune globulin to *Haemophilus influenzae* type b, *Streptococcus pneumoniae* and *Neisseria meningitidis*. Infect Immun 45:248-254

Skov Sørensen U, Szu SC, Henrichsen J (submitted for publication) Association of the cell wall polysaccharide with the capsular polysaccharide of *Streptococcus pneumoniae*.

Sutton A, Schneerson R, Kendall-Morris S, Robbins JB (1982) Differential complement resistance mediates virulence of *Haemophilus influenzae* type b. Infect Immun 35:95-104

Szu SC, Stone AI, Robbins JD, Schneerson R, Robbins JB (1987) Vi capsular polysaccharide-protein conjugates for prevention of typhoid fever. J Exp Med 166:15101524

Tagliabue A, Villa L, de Magistris MT, Romano M, Silvestri S, Boraschi D, Nencioni L (1986) IgA-driven T cell-mediated anti-bacterial immunity in man after live oral Ty 21a vaccine. J Immunol 137:1504-1510

Ward JI, Lum MKW, Hall DB, Silimperi DT, Bendr TR (1986) Invasive *Haemophilus influenzae* type b disease in Alaska: Background epidemiology for a vaccine efficacy trial. J Infec Dis 153:17-27

Wong KH, Barrera O, Sutton A, May J, Hochstein DH, Robbins JD, Robbins JB, Parkman PD, Seligmann Jr EB (1977) Standardization and control of meningococcal vaccines, group A and group C polysaccharides. J Biol Stand 5:197-215

Yount WJ, Dorner M, Kunkel HG, Kabat EA (1968) Studies on human antibodies. VI. Selective variation in subgroup composition and genetic markers. J Exp Med 127:633-646

Protection Induced Against Acute Bronchitis - The Use of Human and Rat Models to Determine Mechanisms of Action of Oral Immunization with *Haemophilus influenzae*

R.L. Clancy[1], F.J. Wallace[1], A.W. Cripps[2], and G.T. Pang[1]

[1]Faculty of Medicine, The University of Newcastle, N.S.W. 2308, and [2]Hunter Immunology Unit, The Royal Newcastle Hospital, Newcastle, N.S.W. 2300, Australia

INTRODUCTION

The demonstrations that both B and T lymphocytes can be activated at one mucosal site, that such activation initiates an intermucosal cell traffic, and that a specific effector response can be detected within the mucosa and secretions of distant mucosal sites (Clancy and Bienenstock 1984; Bienenstock et al. 1983; Bienenstock et al. 1979), provide persuasive support for the concept of a common mucosal system, and a framework for the development of clinically useful vaccines. With respect to bronchus immunity, it has been claimed that optimal stimulation of a local immune response depends upon prior activation of gut-associated lymphoid tissue (Scicchitano et al. 1984), suggesting that a strategy to achieve respiratory tract immunity may logically include ingestion of antigen. However, the development of an effective regimen of oral immunization capable of preventing endobronchitis requires recognition of both the unique characteristics of the pathogenesis of an intralumenal infection, and the restrictions imposed by an inflamed mucosa on the expression of immunity (Clancy and Pucci 1978). The model we have studied has been recurrent acute bronchitis in man, associated with the colonization of damaged airways by non-typable *Haemophilus influenzae* (NTHI). A killed preparation of NTHI ingested by subjects with chronic obstructive lung disease and recurrent acute bronchitis, significantly reduced the frequency of infection (Clancy et al. 1985). The mechanism and specificity of the induced resistance to infection, however, was not clear. Studies on human and animal models are described to provide additional insight into the process of bronchus defense.

HUMAN STUDIES

Forty subjects with a history of recurrent acute bronchitis, but without established chronic obstructive lung disease were studied in a 6 month trial of an orally administered killed NTHI preparation (details of subject population-Table 1). Subjects were randomized into active and placebo groups, with each taking 3 courses of tablets at 0, 28 and 56 days. Each course consisted of 2 tablets each morning for 3 consecutive days. Active tablets each contained 10^{11} killed NTHI organisms. Subjects were reviewed at monthly intervals and at times of acute infection, when throat swabs and sputum samples were taken for microbiological assessment. An acute episode of infection was defined as the production of purulent sputum (or an increase in volume and purulence of sputum if established chronic bronchitis). In the placebo group 11 subjects had a total of 18 episodes of acute bronchitis over the 6 month period, in contrast to 9 subjects having a total of 11 episodes in those taking the active tablets. With respect to the total number of episodes, those taking the active tablets had a 40% reduction in incidence of acute bronchitis ($P < 0.04$ - chi square analysis). Only 25% of sputum samples taken in

Current Topics in Microbiology and Immunology, Vol. 146

acute episodes grew NTHI. Throat swabs taken through the period of study showed a progressive increase in colonization by NTHI only in the placebo group; no difference in colonization by *Haemophilus parainfluenzae* between the two groups was noted (Table 2).

Table 1. Details of subjects in oral immunization study

	Placebo group (20)	Active group (20)
Sex (M:F)	11:9	7:13
Age (yrs.) (range)	47.4 (19-61)	46.3 (27-61)
Current smoker (No. cigarettes/day)	15 22	14 22
Established chronic bronchitis	13	10
No. of acute episodes of bronchitis (2 yrs)	4.25	4.91
FEV_1* (range)	2.67 (1.25-5.0)	2.50 (1.1-4.5)

*Forced expiratory volume in 1 second

Table 2. Influence of ingested *Haemophilus influenzae* on acquisition of Haemophilus species in throat swabs

	March, 1986		June, 1986		September, 1986	
	Placebo	HI	Placebo	HI	Placebo	HI
No. of subjects with *Haemophilus* species (total)	6	7	8	9	7	9
H. influenzae (as % of total growing Haemophilus species)	17	0	25	11	100	11
H. parainfluenzae (as % of total growing *Haemophilus* species	83	57	100	100	100	100

ANIMAL STUDIES

The aim of these studies was to establish a model of chronic bronchitis in which gut-associated lymphoid tissue could be stimulated by NTHI, and clearance of this organism from the respiratory tract could be quantitated as an index of bronchial immunity. Pathogen-free DA rats aged 6 - 8 weeks were used. Clearance was measured as colony forming units (cfu) of NTHI recovered by tracheobronchial lavage, 4 hours after live NTHI were given into the trachea (Toews et al. 1985). Isotype-specific antibody to outer membrane protein (OMP) was determined by ELISA (Yeung et al. 1987), and results expressed as multiples of mean non-immunized values of optical density.

Results of clearance are expressed in Table 3, in terms of immunizing regimen used.

The major observations made were first, that enhanced clearance was only seen when gut-associated lymphoid tissue was initially stimulated, by either direct injection or following oral immunization; second, that while intratracheal immunization alone was not effective, it was necessary following gut priming, for enhanced clearance to be demonstrated; third, systemic immunization could not replace gut priming, and gut priming not followed by local intratracheal stimulation actually prolonged clearance; and fourth, specific antibody did not correlate with enhanced clearance.

IgM anti-OMP antibody was detected (ratio > 2.0) following most immunizing regimens in BAL, but was not detected in saliva (results not shown). Significant enhancement of clearance of bacteria could be demonstrated for at least 40 days following gut priming. When *Klebsiella pneumoniae* was used in the clearance assay following intra Peyer's patch (IPP)/intratracheal (IT) immunizations with NTHI, no enhanced clearance was detected (6.46 ± 0.55).

DISCUSSION

The human studies described oral administration of a killed preparation of NTHI to a population which differed from the previous study (Clancy et al. 1985) in being younger and without clinical evidence of chronic obstructive lung disease. The reduction in incidence of acute bronchitis in those taking the active agent confirmed effectivity of oral immunization in subjects prone to recurrent acute bronchitis, and further supported an aetiologic role for NTHI in acute bronchitis. This trial broadens the target population for oral immunization with killed preparations of NTHI to include younger and fitter subjects. The demonstration of a 40% reduction in incidence in the total number of acute episodes (but not in the number of subjects getting one or more episodes) reflects the lower numbers of episodes associated with NTHI, when compared to subjects with chronic obstructive lung disease (Clancy et al. 1985), and the probability that greater numbers of different aetiological agents cause acute bronchitis in subjects with a relatively normal mucosal defense. Prospective study of colonizing patterns showed that subjects orally immunized with NTHI did not acquire the 'winter peak' seen in the control group. No difference in colonizing pattern, however, was seen with *Haemophilus parainfluenzae*, suggesting that specificity is associated with protection following oral immunization with NTHI.

Studies with the rat model confirmed a need for stimulation of both gut-associated lymphoid tissue, and the bronchus mucosa. Repeated local immunization, or systemic immunization, could not replace gut priming. Prolonged immunity of a specific nature

Table 3

3 (a) Immunization regimen

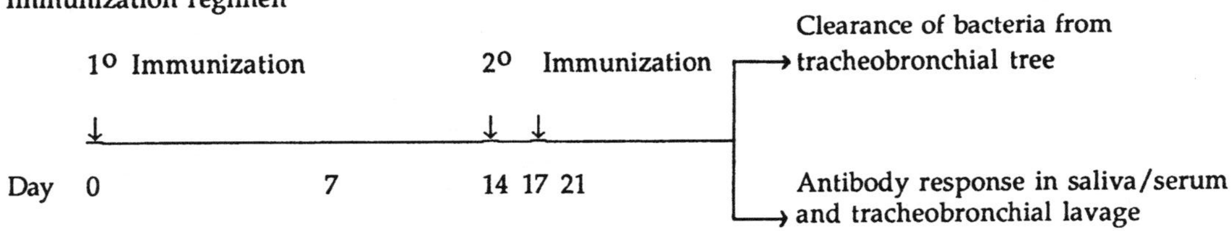

3 (b) Bacterial clearance and antibody response following immunization

Treatment	Number of Animals	Number of viable bacteria (cfu) recovered from tracheobronchial lavage at 4 hrs post-instillation ($\log_{10}$) (mean ± SE)*	Antibody response (+) significant increase ($P < 0.05$) compared with control [saline (IPP)/saline (IT)]		
			Saliva	Serum	Tracheobronchial lavage
Saline (IPP/Saline(IT)	38	6.58 ± 0.16			
HI (IPP)/HI(IT)	6	5.11 ± 0.30*	+	-	+
HI (ORAL)/HI (IT)	5	4.41 ± 0.37*	-	-	-
PBS (IPP)/HI (IT)	5	6.66 ± 0.34	-	-	-
HI (IT)/HI (IT)	5	6.41 ± 0.13	-	+	+
HI (IPP)/PBS (IT)	6	7.78 ± 0.15	-	+	+
HI (S/C)/HI (IT)	5	6.18 ± 0.55	-	+	+

*$P < 0.01$ compared with control group [saline (IPP)/saline (IT)]

HI (IPP) Intra Peyer's patch innoculation 10^{10} cfu

HI (IT) Intratracheal innoculation 5×10^9 cfu at 14 and 17 days post 1° immunization

HI (SC) Subcutaneous injection 5×10^9 cfu

HI (ORAL) Oral feeding 10^9 cfu daily for 14 days

PBS Phosphate buffered saline.

was confirmed, as assessed by enhanced clearance of live NTHI given into the broncho-pulmonary tree.

No specific pattern of IgA antibody to OMP, in upper respiratory tract secretions, was shown to modify the early colonization pattern.

Thus, studies in man and rat complement available data, to confirm a clinical value for oral administration of killed NTHI to subjects prone to recurrent acute bronchitis. Transfer of bronchus immunity with thoracic duct lymphocytes from gut-primed animals (unpublished data) completes a sequence of events describing mechanisms that underpin the protection induced following oral immunization. It remains to be determined whether T and/or B lymphocytes transfer specific immunity. Results from the rat model suggest several approaches to improve current immunization strategies.

Acknowledgment. This study was supported in part by a grant from the Australian Tobacco Foundation.

REFERENCES

Clancy R, Bienenstock J (1984) A framework for immunization strategy. In: Bienenstock J (ed) Immunology of the Lung and Upper Respiratory Tract. Plenum Press, NY, Ch. 9, p 216

Bienenstock J, Befus AD, McDermott MR, Shelagh M, Rosenthal K, Tagliabue A (1983b) The mucosal immunologic network: Compartmentalization of lymphocytes, natural killer cells and mast cells. Ann NY Acad Sci 409:164-170

Bienenstock J, McDermott M, Befus AD (1979) A common mucosal immune system. In: Ogra PL (ed) Immunology of Breast Milk. Raven Press, NY, p 91

Scicchitano R, Husband AJ, Clancy RL (1984) Contribution of gut immunization to the local immune response in the respiratory tract of sheep. Immunology 53:375-384

Clancy RL, Pucci A (1978) Human mucosal lymphocytes - memory for "recall" antigens and non-specific suppressor by T-lymphocytes. Adv Exp Med Biol 107:575-582

Clancy R, Cripps A, Murree-Allen K, Yeung S, Engel M (1985) Protection against acute bronchitis in chronic obstructive lung disease induced by oral immunization with killed *Haemophilus influenzae*. Lancet ii:1395-1397

Toews GB, Hart DA, Hanson EJ (1985) Effect of systemic immunization on pulmonary clearance of *Haemophilus influenzae* Type B. Infect Immun 48:343-349

Yeung S, Pang G, Cripps AW, Clancy RL (1987) Development of an IgA-specific anti-*Haemophilus influenzae* antibody assay for the detection of antibody in human mucosal secretions. J Clin Microbiol 25:667-671

Oral Immunization for the Prevention of Dental Diseases

M.A. Taubman, and D.J. Smith

Department of Immunology, Forsyth Dental Center, 140 Fenway, Boston, MA 02115, USA

Interference with dental caries using immunological stratagems are generally based on three assumptions. The first assumption is that the disease is infectious in nature. The second assumption is that some process in the colonization, growth or pathogenesis of the putative cariogenic bacterium is vulnerable to immunological attack. The third assumption is that products of specific host defense networks can be marshalled at these vulnerable sites for a sufficient time period to interfere with the eventual demineralization of the tooth enamel. We shall discuss the evidence for these assumptions and specifically explore the past experience with, and future potential for immunity induced by the oral route of immunization.

Evidence has implicated a group of acidogenic bacteria, *Streptococcus mutans*, in the etiology of dental caries. These organisms are genetically diverse, falling into at least six serotypes which are collectively referred to as the "mutans" group of streptococci (Coykendall and Gustafson 1986). These organisms (primarily *S. mutans* and *S. sobrinus*) have been identified in human carious lesions (Loesche et al. 1985). They are cariogenic when used to infect animals (Gibbons et al. 1966). The disease is transmissible in animals (Fitzgerald and Keyes 1960) as well as in man, where the mother often is the initial source of infection (Kohler and Bratthall 1977). Reduction or elimination of mutans streptococci from the tooth surface by immunologic (Taubman and Smith 1974) or antibiotic means (Loesche et al. 1977) reduces the incidence of dental caries. Unique biochemical features also implicate these organisms in enamel demineralization. For example, the mutans streptococci produce acid at a high rate and are able to tolerate high sugar, acid and ion concentrations (Van Houte and Russo 1986). Thus, dental caries is an infectious disease and the mutans streptococci are the primary etiologic organisms.

Several steps in the colonization, accumulation, or acid production of the mutans streptococci may be vulnerable to immunological attack. The initial interaction in colonization takes place with the pellicle-coated tooth surface (Clark and Gibbons 1977). The pellicle is primarily composed of salivary components, some of which contain receptors (e.g., alpha galactoside moieties) which will bind mutans streptococci. The components on these streptococci which participate in pellicle-binding appear to be protein, since pepsin diminishes these interactions (Staat et al. 1980). Following the initial attachment phase is a period of accumulation of mutans streptococci. These streptococci produce glucosyltransferases (GTF), which then synthesize water-soluble and water-insoluble glucans from sucrose. Mutans streptococci and certain other oral streptococci express cell surface proteins which serve as binding sites for these extracellular glucans (McCabe et al. 1977; Russell et al. 1983). The interactions between GTF, glucans and cell-bound glucan-binding proteins appear to be fundamental to the accumulation process. This mechanism not only results in an increase in numbers of potentially cariogenic bacteria in plaque, but may also enhance cariogenicity by providing a barrier for diffusion of acid away from the tooth surface.

Current Topics in Microbiology and Immunology, Vol. 146

Thus, there exist several steps at which specific immune mechanisms could theoretically intercept the caries process. Immunization with intact streptococcal cells would elicit immune responses to an array of cell-bound components likely to be involved in several of the phases of colonization, accumulation, or acid production. However, this "shot-gun" approach does not identify the processes which are most vulnerable to immunological control. Furthermore, injection of whole cells may induce unwanted cross-reactive responses with host tissue (Ferretti et al. 1980). For these reasons, considerable effort has been focused on identifying safe but effective antigens from mutans streptococci for use in immunization experiments. Examples of such antigens are the cell wall antigen I/II (Russell and Lehner 1978) which may be involved in the initial attachment phase, and components of the GTF enzyme system which participate in the accumulation phase.

IMMUNOLOGIC PROTECTION AND DENTAL CARIES

The presence of a mucosal immune system suggested that it might be possible to interfere with caries by the stimulation of salivary antibodies to appropriate antigens. Initial studies of the potentially protective effects of the immune system utilized whole cells of mutans streptococci as antigens (Bowen 1969; Taubman and Smith 1974; McGhee et al. 1975). A variety of routes: intravenous, intraductal, subcutaneous, submucosal, or *per os* immunization have conferred protection in various model systems. Some of these experimental approaches with emphasis on the oral route are summarized in Table 1. Most significant were the initial studies in various animal systems which were designed to directly stimulate gut-associated lymphoid tissues (GALT) by ingestion of antigen (usually killed mutans streptococci). These studies demonstrated that consistent induction of secretory IgA antibody to a dental pathogen resulted in reductions in disease and in reductions in cariogenic infection (Michalek et al. 1976; Smith et al. 1980). Importantly, these studies reinforced the previous notion that salivary IgA antibody was the protective principle since virtually no serum (IgG) antibody could be detected (Michalek et al. 1976). It should be noted that similar success was never found in studies of oral immunization of monkeys, although exhaustive investigations involving dose-responsive determinations were never performed (Walker 1981).

A novel modification to this approach has been described in which a recombinant bacterial vector was utilized (Curtiss et al. 1988). This approach involved expression of *S. mutans* antigens on avirulent *Salmonella typhimurium* which attached to, and invaded, Peyer's patches. Although promising, preliminary experiments did not result in sufficient protective antibody to affect dental caries probably because of the relatively sparse production of SpaA protein by these strains (Curtiss et al. 1988).

Another major approach has involved more direct antigen placement on or in oral mucosa, or salivary glands to stimulate local or distant (GALT) elements to provide a salivary IgA response. Direct instillation of *S. mutans* antigen into parotid ducts resulted in elevated antibody production and interference with bacterial colonization (Evans et al. 1975; Emmings et al. 1975). Further efforts involved intramucosal injection of monkeys with *S. mutans* antigen and resulted in serum but no detectable salivary antibody (Bowen et al. 1975). More successful were studies employing topical application of a small *S. mutans* antigen (3.8Kd) and DMSO(50%) to monkey gingival tissues. Immunization of monkeys in this fashion resulted in crevicular IgG, no serum IgG antibody, and whole saliva IgA antibody, reduced caries and proportions of plaque

S. mutans (Lehner et al. 1986). The mechanism of this IgA induction remains unclear although the antibody could have been derived from crevicular fluid. Other studies of topical mucosal application in rodents with whole *S. mutans* did not result in detectable antibody but did appear to reduce colonization (Krasse and Jordan 1977).

Table 1. Experimental immunological approaches

	Approach	References
1.	Active stimulation of GALT.	
	A. Oral in drinking water, feeding, or by intubation - rats, hamsters, monkeys.	Michalek et al. (1976) Smith et al. (1980) Walker (1981)
	B. Intestinal bacteria - *E. coli*, *S. typhimurium* - with GTF or SpaA gene - mice.	Curtiss et al. (1988)
2.	Active local (oral) stimulation of lymphoid tissue in oral mucosa or in salivary glands.	Emmings et al. (1975)
	A. Instillation into parotid glands via ducts - monkeys.	Evans et al. (1975) Walker (1981)
	B. Intramucosal injection - monkeys	Bowen et al. (1975)
	C. Topical gingival application with 3.8 Kd Antigen + DMSO - monkeys.	Lehner et al. (1986)
	D. Topical mucosal application - rodents.	Krasse and Jordan (1977)
3.	Passive oral immunization with monoclonal IgG antibody to antigen I/II.	
	A. Gingival application - monkeys.	Lehner et al. (1985)
4.	Passive oral/GI immunization with polyclonal IgG antibody.	
	A. Ingestion of food supplemented with immune bovine milk - rats.	Michalek et al. (1987)
5.	Passive oral/GI immunization with polyclonal IgA and IgG antibody.	
	A. Suckling of immunized dams - rats.	Michalek and McGhee (1977)

Recently, passive immunization using the oral route has received attention as a means of interfering with dental caries. Usually it is assumed that this interference is with microbial colonization or accumulation steps. One approach has utilized murine monoclonal IgG antibody to streptococcal antigen I/II applied to monkey gingiva (Lehner et al. 1985). This resulted in reduced caries and did not require active immunization with this antigen which may have tissue-reactive properties. Other creative passive immunization approaches in rodents have involved ingestion of food supplemented with immune bovine milk to 4 mutans streptococcal serotypes, and suckling on locally immunized rat dams (Michalek et al. 1977, 1987). These procedures resulted in diminished *S. mutans*, less plaque and reduced caries activity. Thus, while infection with mutans streptococci can lead to dental caries, the immune system seems to regulate the extent of this infection.

IMMUNOLOGIC STUDIES OF MUTANS STREPTOCOCCI IN HUMANS

Most oral immunization studies in humans have utilized intact *S. mutans* or *S. sobrinus* cells as antigen. These studies are summarized in Table 2. The first investigations in this area demonstrated a pronounced secretory immune response both after primary and secondary oral immunization regimes using mutans streptococci administered in capsules (Mestecky et al. 1978). Detection of salivary IgA antibody was facilitated by using subjects with non-detectable antibody prior to immunization. These studies demonstrated induction of a salivary IgA response by stimulation of GALT in humans and reinforced observations in animals (Taubman and Smith 1974) suggesting the presence of an IgA anamnestic response. Subsequent studies using a variety of modest changes in the protocols including topical bacterial administration or a variety of alterations of oral immunization regimens have not shown such pronounced elevated salivary IgA antibody levels (Krasse et al. 1978; Bonta et al. 1979; Cole et al. 1984). In some cases the methodology used was not sufficiently sensitive to detect salivary antibody which shows dramatic fluctuations in antibody content particularly if flow rate varies. However, a unique feature of the studies described above has been the finding of alterations in levels of implanted mutans streptococci (Cole et al. 1984; Krasse et al. 1978; Bonta et al. 1979; Gahnberg and Krasse 1983). More recent studies have suggested that the presence of preexisting indigenous antibody to a particular serotype can result in rapid elimination of that serotype but not of serotype to which there is little or no salivary IgA antibody (Gregory et al. 1985). Further studies of oral immunization with indigenous mutans streptococci resulted in elevated saliva IgA antibody and in decreased indigenous *S. mutans* (Gregory and Filler 1987). Recently the presence of specific IgA antibody-producing mononuclear cells in peripheral blood after oral administration of mutans streptococci in enterically coated gelatin capsules has been demonstrated (Czerkinsky et al. 1987). These predominantly IgA antigen-specific precursor cells peaked in blood 10-12 days after oral immunization and specific IgA antibodies were found in saliva by day 14 to 3 weeks. These experiments provided strong support for the generation of IgA precursor cells after oral immunization followed by emigration of these precursors to selected secretory sites. The one normal individual who did not respond to the oral immunization had high levels of IgA anti-*S. mutans* antibodies and therefore may have been refractory to oral immunization with the same antigen (Czerkinsky et al. 1987). Several important questions of anamnesis, duration, and consistent induction of response by orally administered antigen remain. However, these questions can be addressed by the system described above (Czerkinsky et al. 1987) and in other systems.

Passive oral murine monoclonal antibody to antigen I/II has also been administered to humans (Ma et al. 1987). This antibody resulted in decreased colonization of implanted *S. mutans*.

Table 2. Oral immunization studies in humans

Investigators	Number of subjects	Vaccine	Frequency	Route	Results
Mestecky et al. (1978)	4	Cells (*d*)	14d	Oral	Increased antibody, anamnestic response.
Krasse et al. (1978)	3	Cells (*c*)	8d	Topical	No detectable change in antibody.
Bonta et al. (1979)	4	Cells (*d*)	14d	Oral	No detectable change in antibody, fewer bacteria.
Gahnberg and Krasse (1983)	11	Cells (*d*)	14d	Oral	No change in antibody, fewer bacteria.
Cole et al. (1984)	8	Cells (*d*)	3(7)d	Oral	No change in antibody, fewer bacteria, reductions in levels and duration.
Gregory et al. (1985)	12	Preexisting indigenous antibody to serotype *d* not *c*.	None	Bacterial challenge	Eliminate serotype *d* challenge, not serotype *c*.
Gregory and Filler (1987)	5	Indigenous cells	10d	Oral	Specific salivary IgA antibody and decreased indigenous *S. mutans*.
Czerkinsky et al. (1987)	6	Cells (*c*)	7d	Oral	Specific IgA antibody-producing cells peripheral blood, antibody increase in saliva, tears, fewer bacteria.
Ma et al. (1987)	8	Anti-SA I/II MAb	3X	Passive Oral MAb	Decreased colonization with implanted *S. mutans*.
Smith and Taubman (1987)	25	GTF (*g*) with aluminum phosphate	13X 17d	Oral	Specific IgA antibody in parotid saliva and retardation of indigenous streptococcal reaccumulation.

We investigated the appearance of IgA antibodies in parotid saliva after oral ingestion of GTF by human volunteers (Table 2). A vaccine containing GTF prepared from *S. sobrinus*, suspended in aluminum phosphate, was orally administered, in capsules, to 25 adult males aged 18-36 years (Smith and Taubman 1987). The subjects were assigned to a vaccine or placebo (aluminum phosphate) group based on the IgA antibody activity to *Streptococcus sobrinus* (strain 6715) GTF from parotid saliva samples taken during an initial screening. Antibody levels were compared with those in the parotid saliva of these same individuals taken on six separate occasions prior to the immunization procedure. Subjects gave rise to SIgA antibody in remote salivary gland secretions after oral antigen ingestion. A unique feature of this study was the use of a defined soluble antigen combined with a suspension of aluminum phosphate. Particulate antigen administered by gastric intubation to rodents elicited a higher salivary IgA response than an equivalent dose of the soluble form of the same antigen (Cox and Taubman 1984). These findings may be partially attributed to the enhanced uptake of particulate material by antigen sampling M cells in Peyer's patches (Wolf et al. 1982). Particulate forms of GTF also promote enhanced salivary responses after intragastric immunization of rodents (Ebersole et al. 1983, 1986).

We also found that GTF administration resulted in retardation of reaccumulation of indigenous mutans streptococci after a dental prophylaxis (Smith and Taubman 1987). Presumably these effects were related to the salivary IgA antibody to *S. sobrinus* GTF which will inhibit GTF of other serotypes. Antibody to water-insoluble glucan-synthesizing GTF can markedly inhibit glucan synthesis and accumulation of mutans streptococci *in vitro* (Taubman et al. 1983) and *in vivo*. This effect may result from direct inhibition of GTF-mediated glucan synthesis giving rise to alterations in the nature of dental plaque and restrictions in availability of glucan to binding sites for accumulation of mutans streptococci.

DEVELOPMENT OF SALIVARY IMMUNITY AND MUTANS STREPTOCOCCAL INFECTION

Although young adults were used in the study reported above, the target population for an oral caries vaccine might very well be a much younger group. The most appropriate age for immunization remains a question involving the ontogeny of the secretory immune system and the host age of initial colonization with mutans streptococci. The secretory immune system may not be fully developed at birth (Cooper et al. 1974) but children less than two years of age have secretory antibody to viruses of similar levels to older children (Yanagihara and McIntosh 1980) and demonstrate immunologic memory (Wright et al. 1983). Pakistani children, as young as 2-3 months of age, show saliva IgA antibody to polio virus (Carlsson et al. 1985). Therefore, very young infants demonstrate secretory immune responses.

Streptococcus sanguis is recovered from the dental plaque of nearly all children after the first year of life (Carlsson et al. 1975) and salivary antibody to *S. sanguis* GTF is found in many 2 to 3 year old children (Gahnberg et al. 1985). Mutans streptococci are not consistently recovered in the oral cavity until after the third year of life (Carlsson et al. 1975; Gahnberg et al. 1985), and salivary antibody to *S. mutans* GTF was not detectable in the vast majority of children at 4 years of age (Gahnberg et al. 1985). Therefore, antibody to *S. mutans* GTF is not present when children are colonized with mutans streptococci. Theoretically, immunization with GTF at approximately 12 months of age would encounter a competent immune system. Salivary IgA antibody to GTF could interfere with subsequent colonization and accumulation of mutans streptococci in children. In addition, clinical control of the level of mutans streptococcal infection in mothers has been shown to limit, or virtually eliminate, the infectious challenge (Kohler et al. 1983). The reduction of challenge coupled with the presence of appropriate salivary antibody before permanent colonization by mutans streptococci should ultimately result in a significant caries reduction. Employment of this strategy promises to provide an effective intervention in the dental infectious process.

Acknowledgments. The research reported in this manuscript was supported by grants DE-04733 and DE-06153 from the NIDR and by Lederle Laboratories. We thank Donna Freedman for secretarial assistance.

REFERENCES

Bonta CY, Linzer R, Emmings F, Evans RT, Genco RJ (1979) Human oral infectivity and immunization studies with *Streptococcus mutans* strain B13. J Dent Res 58:143

Bowen WH, Cohen B, Colman G (1975) Immunization against dental caries. Brit Dent J 139:45-58

Bowen WH (1969) A vaccine against dental caries. A pilot experiment in monkeys (*Macaca irus*). Brit Dent J 126:159-160

Carlsson J, Grahnen H, Jonsson G (1975) Lactobacilli and streptococci in the mouths of children. Caries Res 9:333-339

Carlsson B, Zaman S, Mellander L, Jalil F, Hanson LÅ (1985) Secretory and serum immunoglobulin class-specific antibodies to poliovirus after vaccination. J Infect Dis 152:1238-1244

Clark WB, Gibbons RJ (1977) Influence of salivary components and extracellular polysaccharide synthesis from sucrose on the attachment of *Streptococcus mutans* 6715 to hydroxyapatite surfaces. Infect Immun 18:514-523

Cole MF, Emilson CG, Hsu SD, Li S-H, Bowen WH (1984) Effect of peroral immunization of humans with *Streptococcus mutans* on induction of salivary and serum antibodies and inhibition of experimental infection. Infect Immun 46:703-709

Cooper MD, Kincade PW, Bockman DE, Lawton AR (1974) Origin, distribution and differentiation of IgA-producing cells. Adv Exp Med Biol 45:13-22

Cox DS, Taubman MA (1984) Oral induction of the secretory antibody response by soluble and particulate antigens. Arch Allergy Appl Immunol 75:126-131

Coykendall AL, Gustafson KB (1986) Taxonomy of *Streptococcus mutans.* In: Hamada S, Michalek SM, Kiyono H, Menaker L, McGhee JR (eds) Molecular Microbiology and Immunobiology of *Streptococcus mutans,* Elsevier, Amsterdam, p 21

Curtiss R III, Goldschmidt R, Kelly S, Lyons M, Michalek SM, Pastien R, Stein S (1988) Recombinant avirulent *Salmonella* for oral immunization to induce mucosal immunity to bacterial pathogens. In: Kohler H, LoVerde PT (eds) Vaccines New Concepts and Developments, Longman and John Wiley, NY p 261

Czerkinsky C, Prince SJ, Michalek SM, Jackson S, Russell MW, Moldoveanu Z, McGhee JR, Mestecky J (1987) IgA antibody-producing cells in peripheral blood after antigen ingestion: Evidence for a common mucosal immune system in humans. Proc Natl Acad Sci USA 84:2449-2453

Ebersole JL, Taubman MA, Smith DJ (1983) Adjuvants, glucosyltransferases and caries vaccine. In: Doyle RJ, Ciardi JE (eds) Glucosyltransferase, Glucans, Sucrose and Dental Caries. Special Supplement Chemical Sense, Washington, DC, p 241

Ebersole JL, Taubman MA, Smith DJ, Frey DE (1986) Regulation of salivary SIgA antibody. In: Hamada S, Michalek SM, Kiyono H, Menaker L, McGhee JR (eds) Molecular Microbiology and Immunology of *Streptococcus mutans*, Elsevier, Amsterdam, p 329

Emmings FG, Evans RT, Genco RJ (1975) Antibody response in the parotid fluid and serum of irus monkeys (*Macaca fascicularis*) after local immunization with *Streptococcus mutans*. Infect Immun 12:281-292

Evans RT, Emmings FG, Genco RJ (1975) Prevention of *Streptococcus mutans* infection of monkeys (*Macaca fascicularis*). Infect Immun 12:293-302

Ferretti JJ, Shea C, Humphrey MW (1980) Cross-reactivity of *Streptococcus mutans* antigens and human heart tissue. Infect Immun 30:69-73

Fitzgerald RJ, Keyes PH (1960) Demonstration of the etiologic role of streptococci in experimental caries in the hamster. J Am Dent Assoc 61:9-19

Gahnberg L, Krasse B (1983) Salivary immunoglobulin A antibodies and recovery from challenge of *Streptococcus mutans* after oral administration of *Streptococcus mutans* vaccine in humans. Infect Immun 39:514-519

Gahnberg L, Smith DJ, Taubman MA, Ebersole JL (1985) Salivary IgA antibody to glucosyltransferase of oral microbial origin in children. Arch Oral Biol 30:551-556

Gibbons RJ, Berman KS, Knoettnor P, Kapsimalis B (1966) Dental caries and alveolar bone loss in gnotobiotic rats infected with capsule forming streptococci of human origin. Arch Oral Biol 11:549-560

Gregory RL, Filler SJ (1987) Protective secretory immunoglobulin A antibodies in humans following oral immunization with *Streptococcus mutans*. Infect Immun 55:2409-2415

Gregory RL, Michalek SM, Filler SJ, Mestecky J, McGhee JR (1985) Prevention of *Streptococcus mutans* colonization by salivary IgA antibodies. J Clin Immunol 5:55-61

Kohler B, Bratthall D (1977) Intrafamilial levels of *Streptococcus mutans*: A comparison between mother and child. J Dent Res Spec Issue A 56:A157

Kohler B, Bratthall D, Krasse B (1983) Preventive measures in mothers influence the establishment of the bacterium *Streptococcus mutans* in their infants. Arch Oral Biol 28:225-231

Krasse B, Jordan HV (1977) Effect of orally applied vaccines on oral colonization by *Streptococcus mutans* in rodents. Arch Oral Biol 22:479-484

Krasse B, Gahnberg L, Bratthall D (1978) Antibodies reacting with *Streptococcus mutans* in secretions from minor salivary glands in humans. Adv Exp Med Biol 107:349-354

Lehner T, Caldwell J, Smith R (1985) Local passive immunization by monoclonal antibodies against streptococcal antigen I/II in the prevention of dental caries. Infect Immun 50:796-799

Lehner T, Mehlert A, Caldwell J (1986) Local active gingival immunization by a 3,800-molecular-weight streptococcal antigen in protection against dental caries. Infect Immun 52:682-687

Loesche WJ, Bradbury DR, Woolfork MP (1977) Reduction of dental decay in rampant caries individuals following short-term kanamycin treatment. J Dent Res 56:254-265

Loesche WJ, Eklund S, Earnest R, Burt B (1985) Longitudinal investigation of bacteriology of human fissure decay: Epidemiological studies in molars shortly after eruption. Infect Immun 46:765-772

Ma JK-C, Smith R, Lehner T (1987) Use of monoclonal antibodies in local passive immunization to prevent colonization of human teeth by *Streptococcus mutans*. Infect Immun 55:1274-1278

McCabe MM, Hamelik RM, Smith EE (1977) Purification of dextran-binding protein from cariogenic *Streptococcus mutans*. Biochem Biophys Res Commun 87:273-278

McGhee JR, Michalek SM, Webb J, Navia JM, Rahman AFR, Legler DW (1975) Effective immunity to dental caries: Protection of gnotobiotic rats by local immunization with *Steptococcus mutans*. J Immunol 16:300-305

Mestecky J, McGhee JR, Arnold RR, Michalek SM, Prince SJ, Babb JL (1978) Selective induction of an immune response in human external secretions by ingestion of bacterial antigen. J Clin Invest 61:731-737

Michalek SM, McGhee JR (1977) Effective immunity to dental caries: Passive transfer to rats of antibodies to *Streptococcus mutans* elicits protection. Infect Immun 17:644-650

Michalek SM, McGhee JR, Mestecky J, Arnold RR, Bozzo L (1976) Ingestion of *Streptococcus mutans* induces secretory IgA and caries immunity. Science 192:128-1240

Michalek SM, Gregory RL, Harmon CC, Katz J, Richardson GJ, Hilton T, Filler SJ, McGhee JR (1987) Protection of gnotobiotic rats against dental caries by passive immunization with bovine milk antibodies to *Streptococcus mutans*. Infect Immun 55:2341-2347

Russell MW, Lehner T (1978) Characterization of antigens extracted from cells and culture fluids of *Streptococcus mutans* serotype *c*. Arch Oral Biol 23:7-15

Russell RR, Donald AC, Douglas CW (1983) Fructosyltransferase activity of a glucan-binding protein from *Streptococcus mutans*. J Gen Microbiol 129:3242-3250

Smith DJ, Taubman MA (1987) Oral immunization of humans with *Streptococcus sobrinus* glucosyltransferase. Infect Immun 55:2562-2569

Smith DJ, Taubman MA, Ebersole JL (1980) Local and systemic antibody response to oral administration of glucosyltransferase antigen complex. Infect Immun 28:441-450

Staat RH, Langley SD, Doyle RJ (1980) *Streptococcus mutans* adherence: Presumptive evidence for protein-mediated attachment followed by glucan-dependent cellular accumulation. Infect Immun 27:675-681

Taubman MA, Smith DJ (1974) Effects of local immunization with *Streptococcus mutans* on induction of salivary immunoglobulin A antibody and experimental dental caries in rats. Infect Immun 9:1079-1091

Taubman MA, Smith DJ, Ebersole JL, Hillman JD (1983) Protective aspects of immune response to glucosyltransferase in relation to dental caries. In: Doyle RJ, Ciardi JE (eds) Glucosyltransferase, Glucans, Sucrose and Dental Caries, Special Supplement, Chem Senses, Washington DC, p 249

van Houte J, Russo J (1986) Factors influencing the cariogenicity of *Streptococcus mutans*. In: Hamada S, Michalek SM, Kiyono H, Menaker L, McGhee JR (eds) Molecular Microbiology and Immunology of *Streptococcus mutans*, Elsevier, Amsterdam, p 157

Walker J (1981) Antibody responses of monkeys to oral and local immunization with *Streptococcus mutans*. Infect Immun 31:61-70

Wolf JL, Rubin DH, Finberg R, Kauffman RS, Sharpe AH, Trier JS, Fields BN (1982) Intestinal M cells: A pathway of entry for reovirus into the host. Science 212:471-472

Wright PF, Murphy BR, Kervina M, Lawrence EM, Phelan MA, Karzon DT (1983) Secretory immunological response after intranasal inactivated influenza A virus vaccination: Evidence for immunoglobulin A memory. Infect Immun 40:1092-1095

Yanagihara R, McIntosh K (1980) Secretory immunological response in infants and children to parainfluenza virus types 1 and 2. Infect Immun 30:23-28

Oral Immunization Against Cholera

J. Holmgren[1], J. Clemens[2], D.A. Sack[2], J. Sanchez[1], and A.-M. Svennerholm[1]

[1]Department of Medical Microbiology, University of Göteborg, S-413 46 Göteborg, Sweden; and [2]International Centre for Diarrhoeal Disease Research, Bangladesh, Dhaka, Bangladesh

INTRODUCTION

Cholera is an important cause of morbidity and mortality in many developing countries. Convalescents from this disease develop a substantial but not complete immunity against new attacks of cholera (Glass et al. 1982). Although cholera can be effectively treated by oral and/or intravenous rehydration, development of a vaccine that is capable of preventing cases and reducing the severity of cholera would provide the best immediate possibility to control the disease. If this vaccine could also provide protection against diarrhea caused by heat-labile enterotoxin (LT)-producing *Escherichia coli* its importance as a preventive health measure would be even more significant (Holmgren and Svennerholm 1985).

Because parenteral vaccination against cholera has yielded only modest and short-term protection (usually less than 50% protection for less than 6 months), attention has turned to development of oral vaccines that stimulate intestinal immunity more efficiently. Furthermore, since natural protection from cholera appears to result from local intestinal antibacterial and antitoxic antibodies, which act synergistically, special efforts have been made to develop vaccines that could effectively stimulate these antibodies (Holmgren and Svennerholm 1983).

Attenuated live organisms are appealing as oral cholera vaccine candidates because of their ability to colonize the intestine and to stimulate an immune response in a manner analogous to natural cholera infection (Levine et al. 1983). However, the live attenuated cholera strains developed thus far have caused diarrhea when fed to human volunteers which has precluded their further testing as potential vaccines. In contrast, an oral cholera vaccine containing killed bacteria and purified B subunit "toxoid" has been developed which does not cause any side-effects and yet gives rise to long-lasting protective immunity as evident from a large-scale field trial in a highly cholera endemic area.

THE ORAL B-SUBUNIT - WHOLE CELL CHOLERA VACCINE

The vaccine, referred to as the "oral B subunit/whole cell vaccine" (B/WCV), consists of purified B-subunit from cholera toxin and formalin- or heat-inactivated classical and El Tor cholera vibrios of the Inaba and Ogawa serotypes (Svennerholm and Holmgren 1986). Because of the acid sensitivity of particularly the B subunit component, the vaccine has been administered after and/or together with a sodium bicarbonate/citric acid buffer solution to ensure adequate neutralization of stomach acidity for preservation of the vaccine when passing through the stomach.

Current Topics in Microbiology and Immunology, Vol. 146

The development of the oral B/WCV originated from our basic studies in the 1970s of the mechanisms of disease and immunity in cholera. This work has since been followed by 10 years of applied development and systematic testing of the vaccine in collaboration with the International Centre for Diarrhoeal Disease Research, Bangladesh (ICDDR,B); Institut Merieux, France; The National Bacteriological Laboratory, Sweden; and The University of Maryland, USA.

The B/WCV vaccine has proved to be completely safe when tested in large numbers of Swedish, Bangladeshi and American volunteers. It has also been found to stimulate strong IgA mucosal immune responses against cholera toxin as well as cholera vibrios in the gut of these volunteers. The gut mucosal IgA antitoxin and anti-LPS antibody formation and local immunologic memory evoked by 2 doses of the vaccine closely resembled the immune response seen in convalescents from cholera (Svennerholm et al. 1984). In American volunteers who received 3 oral immunizations with either the B/WCV combination vaccine or its WCV component alone, the vaccine-induced immune response proved to be associated with substantial protection against challenge with live cholera vibrios, and protection was complete against clinically severe diarrhea (Black et al. 1987).

FIELD TRIAL IN BANGLADESH

The ultimate test of the level of protection afforded by cholera vaccines is a field trial for efficacy in a cholera endemic area. The promising findings obtained with particularly the combined B/WCV but also with the whole cell constituents of this vaccine alone (WCV) prompted the International Centre for Diarrhoeal Disease Research, Bangladesh (ICDDR,B) in collaboration with the government of Bangladesh and the World Health Organization, to mount a randomized, double-blind field trial of the two oral vaccines in rural Bangladesh.

Before starting the main trial a pretest was undertaken in which 1257 Bangladeshi villagers were randomized to receive 3 doses of one of the two vaccines, a heat-killed *Escherichia coli* K12 strain placebo, or distilled water. The lots of vaccines and placebo for the main trial caused no detectable side effects and elicited the expected rises in serum vibriocidal and anti-cholera-toxin antibodies (Clemens et al. 1987).

The protective efficacy of the oral B/WCV and WCV vaccines was then assessed in 84,000 Bangladeshi children aged 2-15 years and women aged over 15 years (Clemens et al. 1986). Among these, 63,498 persons received three doses and the remaining 20,502 one or two doses of B/WCV, WCV alone or K12 placebo in a randomized, double-blinded fashion. The field trial was conducted in the Matlab field study area of the ICDDR,B. Matlab lies in the Ganges delta 45 km southeast of Dhaka, and ca. 190,000 persons currently reside in this area. Cholera is endemic in Matlab with a median yearly incidence of 1.5 hospital cases/1000 general population. Vaccination proceeded in three 6-week rounds beginning in January, 1985. The vaccines and placebo were administered by 69 vaccination teams. Surveillance for diarrhea was then maintained at the three diarrheal treatment centers serving the Matlab population. Uniform information about the clinical status at the time of presentation and during the subsequent clinical course was entered onto data forms for computer entry. A stool specimen or rectal swab was obtained from each patient and was cultured to identify *V.*

cholerae O1 and to determine the biotype and serotype of each isolate as well as other known enteric pathogens including enterotoxigenic *E. coli* (ETEC).

Protection Against Cholera

Table 1 presents the overall results of the trial as concerns protection against cholera during two years of follow-up. For the initial 6 month surveillance period the combined B/WCV vaccine conferred a high degree (85%) of protection against cholera ($p<0.001$ one-tailed). The WCV only vaccine also afforded significant, although less impressive protection (protective efficacy 58%; $p<0.01$ one-tailed). The group vaccinated with B/WCV had 64% fewer cases of cholera than the group receiving the WCV only vaccine ($p=0.04$ one-tailed). Furthermore, subgroup analyses showed that the protective efficacy for both vaccines during the first 6 months was consistent for children aged 2-10 years and older persons (92% protective efficacy in children for the combined vaccine) (Clemens et al. 1986).

Table 1. Efficacy of oral B subunit/whole-cell cholera vaccine, or whole-cell vaccine alone (Data from field trial, Bangladesh; Clemens et al. 1986; 1988a)

Period of surveillance and age group	Vaccine efficacy for cases of cholera presenting for treatment (%) [a] Whole-cell/B subunit	Whole-cell only
First 6 months, all ages	85	58
First 12 months, all ages	62 (74)	53 (16)
Age ≥ 6 years	76	68
Age 2-5 years	38	31
First 24 months, all ages	60	58

[a] 3 doses of vaccine or placebo were given at intervals of 1 month; values in parentheses are protection in recipients of only 2 doses.

After one complete year of surveillance, during which the population experienced an exceptionally high incidence of cholera with two major outbreaks in the later part of the year the B/WCV vaccine provided 62% protection and the WCV alone 53% protection after 3 doses, as compared to 74% for B/WCV and 16% for WCV alone after 2 doses (Clemens et al.1988a). High-grade and sustained protection was observed in persons vaccinated at ages over five years: 76% for B/WCV and 68% for WCV; in younger persons protection declined during the second as compared to the first 6 month period. Both vaccines conferred equivalent protection against cholera associated with life-threatening dehydration and against cholera associated with less severe manifestations, and both conferred somewhat greater efficacy against classical than against El Tor cholera. Protection was markedly greater for B/WCV recipients than for WCV recipients during the initial 8 months of observation, but was similar thereafter.

Preliminary data is now available on protection after a second year of follow-up, during which the cholera incidence rate in the field area was also unusually high with much cholera occurring especially during the first half of the second year. During this second year of observation the pattern of protection from the first year has remained essentially stable, giving an overall protection of approx. 60% for both vaccines. However, it is of interest to note that protection for the last 6 months of the two years was 80% for the B/WCV recipients, which suggests that protection may last for considerably longer than two years.

Protection Against Enterotoxigenic *E. Coli* Diarrhea

Because of the antigenic similarity of the B-subunits of cholera toxin and of the heat-labile enterotoxin (LT) of enterotoxigenic *E. coli* (ETEC), the protective efficacy of the oral B/WCV cholera vaccine against LT-producing ETEC diarrhea was assessed within the framework of the randomized, double-blinded field trial of the oral cholera vaccines in Bangladesh (Clemens et al. 1988b). The efficacy of the B/WCV was evaluated by comparing the occurrence of LT-ETEC in persons who had received this vaccine with that in persons who had taken the cholera WCV component alone. During the first 3 months of follow-up after 2 or 3 doses the B/WCV gave a high level of protection (Table 2). The protective efficacy was similar for disease caused by ETEC that produced LT-only and for strains that produced (*in vitro*) both LT and heat-stable toxin (ST). This suggests that in patients infected with LT/ST ETEC, LT played the main (or only) pathogenic role. A notably greater protective effect (86%) was observed against LT-ETEC associated with potentially life-threatening dehydration than against less severe illness (54%). This difference was more pronounced after longer periods of observation; thus the one-year protection against severe disease was 50% as compared with 21% for ETEC diarrhea of any severity. From these results we conclude that the B-subunit component of the B/WCV cholera vaccine conferred high-grade cross-protection against LT-ETEC diarrhea.

Table 2. Cross-protection by oral B subunit/whole cell cholera vaccine against LT or LT/ST *E. coli* diarrhea: Results of a large-scale field trial (Clemens et al. 1988b)

E. coli diarrhea	Protective efficacy after 2 or 3 doses (%) [a]	
	0-3 months	0-12 months
Severe	86	50
Non-severe	54	--
Total	67	21

[a] Comparison between disease incidence among 24,770 recipients of B/WCV versus 24,842 recipients of placebo (WCV without B subunit) vaccines.

Reduction of Overall Morbidity and Mortality

Finally, and of great potential importance, recent data from the trial indicate that the oral cholera vaccines also had a substantial impact on overall morbidity in diarrheal disease (Clemens et al. 1988c). During the first year of follow-up, there was a 26% reduction of all visits for treatment of diarrhea in the B/WCV group and a 22%

reduction in the WCV only group. The impact was particularly great on severely dehydrating, life-threatening diarrhea which was reduced by at least 48% with the B/WCV as compared with 32% with the WCV only. The greater efficacy of the combined vaccine against severe watery diarrhea may have been due both to the added effect of the B subunit against cholera and to protection by the B subunit against LT-producing ETEC and possibly also other bacterial pathogens producing cholera-like enterotoxins as well.

Field surveillance of deaths from all causes among trial participants revealed that overall mortality among women above the age of 15 years was reduced by 45% and 33% for the first year in those given the combined vaccine or WCV, respectively (Clemens et al. 1988c). This effect was due largely to reduced mortality during the cholera "epidemic season", during which mortality rates in this age group increased almost 3-fold in the placebo group. Deaths, which according to "verbal autopsy" were reported to be associated with non-dysentery diarrhea, were reduced by approximately 80%. These results suggest that the vaccines prevented many diarrhea-associated deaths which would otherwise have occurred among persons who did not seek care at local treatment centers.

We conclude that the killed oral cholera vaccines had a marked overall effect on admissions for life-threatening diarrhea and may have reduced the overall risk of death in adult women.

IMPROVEMENTS IN VACCINE PRODUCTION TECHNOLOGIES

The procedures for preparation and characterization of the B/WCV vaccine used in the field trial are well defined. The B subunit component is prepared from the culture filtrate of fermentor-grown *V. cholerae* 569B bacteria by affinity chromatography on a specific toxin-binding Spherosil-GM1 ganglioside column followed by gel filtration on a Sephadex column in acid buffer (Tayot et al. 1981). By this procedure approx. 50 g (50,000 doses) of highly purified B subunit can be obtained from each 1000-liter culture. The heat-killed organisms in the whole cell vaccine are prepared by standard procedures for cholera vaccines using established Inaba and Ogawa cholera vaccine strains. The formalin-killed organisms that are also included in the whole cell vaccine represent an El Tor and a classical *V. cholerae* strain that have been selected because they are particularly good and stable producers of the main cell-associated hemagglutinins associated with El Tor and classical *V. cholerae* organisms, respectively. These organisms are grown and killed using procedures found to optimize the expression and preservation of these heat-labile hemagglutinins as well as other surface protein antigens (A-M Svennerholm et al. to be published).

A limiting factor today in the production of the oral B/WCV cholera vaccine, which is especially important with regard to the transfer of production technology to developing countries for their local production of the complete vaccine, is the need to produce the B-subunit from cholera toxin. This step requires sophisticated biochemical procedures as well as stringent purification and toxicity assessment criteria to insure the preparation of a safe, non-toxic vaccine.

To overcome this problem we have, using recombinant DNA technology, tried to place the structural gene for the cholera toxin B-subunit (without any A subunit) under the control of a very strong promoter in a plasmid. Our results now indicate that we have succeeded in generating such a plasmid and expressing it in cholera toxin gene-deleted *V. cholerae* organisms to yield production of high levels of cholera B-subunit in the absence of any toxin (Sanchez and Holmgren 1988).

We decided to place the cholera B subunit (CTB) gene under the control of the strong *tac* promoter in the wide-host-range plasmid pMMMB68. This approach was chosen because the *E. coli* LTB gene has recently been inserted behind the *tac* promoter in this plasmid (Sandkvist et al. 1987), and when this plasmid was expressed in the cholera toxin deleted *V. cholerae* strain JBK70 (El Tor, Inaba) we obtained up to 50-100 µg/ml of LTB subunit in the culture supernatant (unpublished). We prepared a DNA fragment carrying the CTB gene and ligated its 5' end to the 3' end of the DNA for the LTB leader, thus giving an in-frame fusion of the leader peptide of LTB to the CTB mature protein under the control of the *tac* promoter. We assumed that the presence of the foreign leader peptide would not interfere with the recovery of CTB since this part of the protein is normally removed by *E. coli* or *V. cholerae* hosts during protein secretion. Two closely related plasmids, pJS161 and pJS162 were generated by this method. The *tac* promoter is inducible with isopropylthiogalactopyranoside (IPTG). When IPTG was supplied to the medium of *E. coli* K12 organisms harboring the pJS161 or pJS162 plasmids, it was found that the plasmids directed the synthesis of cholera B subunit at higher levels than previously seen in recombinant *E. coli* strains. The authenticity of the recombinant cholera B subunit was ascertained using monoclonal antibodies which completely lack cross-reactivity with the LTB protein. When the plasmids were then transferred to *V. cholerae* JBK70, and the strain was grown in the presence of IPTG, levels of 50-100 µg/ml of secreted recombinant cholera B subunit were regularly produced (Table 3); this is approx. 100 times more than for other described recombinant cholera strains.

Table 3. Expression of recombinant cholera B subunit by the *tac* promoter in *V. cholerae* (Sanchez et al. to be published)

pJS162 in	Biotype	Serotype	µg/ml of CTB
V. cholerae			
JBK70	El Tor	Inaba	94
JS1395 (CVD101 rif^{r})	Classical	Ogawa	54
JS1569 (CVD103 rif^{r})	Classical	Inaba	75

Based on these results, it is now possible to develop a simplified procedure for the production of the B/WCV cell cholera vaccine based on two "B-only" plasmid-carrying *V. cholerae* strains representing the different serotypes as well as biotypes. The JBK70 *V. cholerae* strain is of El Tor biotype and of Inaba serotype. Therefore we have attempted to transfer the pJS162 plasmid into toxin-deleted *V. cholerae* organisms of classical biotype. For this purpose we first isolated rifampicin-resistant strains (for counter-selection). Subunit-deleted classical *V. cholerae* mutants of both Inaba and Ogawa serotypes have been selected. Our preliminary results with these strains (Table 3) indicate that these classical organisms can also produce high levels of cholera B-subunit (Sanchez and Holmgren 1988, to be published). Thus, we may already have available the necessary vaccine strains to enable the proposed simplified preparation of the oral B/WCV cholera vaccine, although much work remains before this approach will yield a practical vaccine that can be produced locally by developing countries where the cholera vaccine is needed.

REFERENCES

Black RE, Levine MM, Clements ML, Young CR, Svennerholm A-M, Holmgren J (1987) Protective efficacy in man of killed whole vibrio oral cholera vaccine with and without the B subunit of cholera toxin. Infect Immun 55:1116-1120

Clemens JD, Sack DA, Harris JR, Chakraborty J, Khan MR, Stanton BF, Kay BA, Khan MU, Yunus MD, Atkinson W, Svennerholm A-M, Holmgren J (1986) Field trial of oral cholera vaccines in Bangladesh. Lancet ii: 124-127

Clemens J, Stanton BF, Chakraborty J, Sack D, Khan MR, Huda S, Ahmed F, Harris JR, Yunus M, Khan MU, Svennerholm A-M, Jertborn M, Holmgren J (1987) B subunit-whole cell and whole cell only killed oral vaccines against cholera: Studies on reactogenicity and immunogenicity. J Infect Dis 155:79-85

Clemens JD, Harris JR, Sack DA, Chakraborty J, Ahmed F, Stanton BF, Khan MU, Kay BA, Huda N, Khan MR, Yunus, M, Rao MR, Svennerholm A-M, Holmgren J (1988) Field trial of oral cholera vaccines in Bangladesh: Results of one year of follow-up. J Infect Dis 158:60-69

Clemens JD, Sack DA, Harris JR, Chakraborty J, Neogy PK, Stanton B, Huda N, Khan MU, Kay BA, Khan MR, Ansaruzzaman M, Yunus M, Rao MR, Svennerholm A-M, Holmgren J (in press) Cross-protection by B subunit-whole cell cholera vaccine against diarrhoea associated with heat labile toxin-producing enterotoxigenic *Escherichia coli*: Results of a large-scale field trial. J Infect Dis

Clemens JD, Sack DA, Harris JR, Chakraborty J, Khan MR, Stanton B, Ali M, Huda N, Yunus M, Kay BA, Khan MU, Rao MR, Svennerholm A-M, Holmgren J (1988) Impact of B subunit killed whole-cell and killed whole-cell-only oral vaccines against cholera upon treated diarrhoeal illness and mortality in an area endemic for cholera. Lancet i:1375-1378

Glass RI, Becker S, Huq IM, Stoll BJ, Kahn MV, Merson MH (1982) Endemic cholera in rural Bangladesh, 1966-1980. Am J Epidemiol 116:959-970

Holmgren J, Svennerholm A-M (1983) Cholera and the immune response. Progr Allergy 33:106-119

Holmgren J, Svennerholm A-M (1985) Vaccine development for the control of cholera and related toxin-induced diarrheal diseases. CIBA Found Symp 112, Pitman, London, p 242

Levine MM, Kaper JB, Black RE, Clements ML (1983) New knowledge on pathogenesis of bacterial enteric infections as applied to vaccine development. Microbiol Rev 47:510-550

Sanchez J, Holmgren J (in press) Recombinant system for overexpression of cholera toxin B-subunit in *Vibrio cholerae* as a basis for vaccine development. Proc Natl Acad Sci USA

Sandkvist M, Hirst TR, Bagdasarian M (1987) Alterations at the carboxyl terminus change assembly and secretion properties of the B subunit of *Escherichia coli* heat-labile enterotoxin. J Bacteriol 169:4570-4576

Svennerholm A-M, Holmgren J (1986) Oral combined - whole cell cholera vaccine. In: Holmgren J, Lindberg A, Möllby R (eds) Development of vaccines and drugs against diarrhea, 11th Nobel Conference, Stockholm 1985, Studentlitteratur Lund, p 33

Svennerholm A-M, Jertborn M, Gothefors L, Karim AMMM, Sack DA, Holmgren J (1984) Comparison of mucosal antitoxic and antibacterial immune responses after clinical cholera and after oral immunization with a combined B subunit-whole cell vaccine. J Infect Dis 149:884-893

Tayot JL, Holmgren J, Svennerholm L, Lindblad M, Tardy M (1981) Receptor-specific large-scale purification of cholera toxin on silica beads derivatized with lyso GM1 ganglioside. Eur J Biochem 113:249-258

Oral Shigella Vaccines

T.L. Hale, and S.B. Formal

Department of Enteric Infections, Walter Reed Army Institute of Research, Washington, D.C. 20307-5100, USA

INTRODUCTION

Shigellae cause intestinal disease by invading the colonic epithelium. Intracellular multiplication and intercellular spread of these organisms results in mucosal lesions and inflammation of the lamina propria (LaBrec et al. 1964). These lesions are usually confined to the colon, and they are responsible for the bloody, mucoid stools characteristic of bacillary dysentery. Naturally acquired shigella infections occur only in primates, and challenge studies have shown that the ID_{50} of *Shigella flexneri 2a* is approximately 5 X 10^3 colony forming units (CFU) (DuPont et al. 1969). Previous infection provides about 75% protection against a subsequent challenge with the ID_{50} of the homologous strain (Dupont et al. 1972); therefore, immunization against shigellosis is a theoretical possibility.

Early attempts to develop safe and effective shigella vaccines involved parenteral injection of whole bacterial cells. Although these vaccines elicited high titers of circulating antibody, they were unable to protect monkeys in oral shigella challenge experiments (Formal et al. 1967) or to protect humans in field conditions (Higgins et al. 1955). Recently, parenteral immunization of monkeys with ribosomal vaccine preparations from *S. sonnei* have yielded more promising results (Dzikidze et al. 1985). Nonetheless, most research and development within the last twenty years has concentrated on living oral vaccines. The characteristics of these vaccines are summarized in Table 1.

MUTANT ATTENUATED VACCINES

Three types of spontaneously arising avirulent mutants of *S. flexnuri 2a* were isolated in the early sixties. One of these, 24570, is an opaque colonial variant which no longer expresses the invasive phenotype (LaBrec et al. 1964). Several massive doses of this mutant effectively protected monkeys from challenge with the virulent 2457T parent strain, but 10^{10} CFU of 24570 caused dysentery in 34% of human volunteers (DuPont et al. 1972). Since 24570 isolates from ill volunteers had reverted to the invasive phenotype, this vaccine was deemed too unstable for use in human populations.

The discovery of a family of large plasmids which are necessary for expression of the invasive phenotype in *S. flexneri* (Sansonetti et al. 1982), has allowed the reinterpretaton of some of the earlier work on the 24570 strain. It is now clear, that the 24570 strain retains the 140 mega Dalton (mDal) plasmid (Kopecko et al. 1979), but it no longer expresses the invasion plasmid antigen (*ipa*) proteins (Buysse et al. 1987) which are associated with the invasive phenotype (Hale et al. 1985). Since the *ipa* genes are still present in 24570, a population of invasive revertants was apparently selected when

Current Topics in Microbiology and Immunology, Vol. 146

Table 1. *Shigella* vaccines

Immunizing Agent	Genotype	Phenotype	Efficacy	Safety
1. Mutant attenuated *Shigella flexneri*				
24570	Unidentified mutation(s) down relate 140 mDal invasion plasmid	Noninvasive	Yes (in monkeys)	No (reverts to virulence)
T32 Istrati	Unidentified mutation(s)	Noninvasive (?)	Yes (multiple doses)	Yes
Streptomycin-dependent (SmD)	Chromosomal mutation to SmD and loss	Noninvasive (?) nonreplicating	Yes (multiple doses)	Yes (reverts to Sm independence)
2. Attenuated *Shigella flexneri* hybrids				
M22-18X16	*S. flexneri* 2a with *xyl-rha* genes from *Escherichia coli* K-12	Invasive but nonreplicating	Yes (single dose)	No (insufficient attenuation)
Sf1-114	*S. flexneri* Y with Tn10 insertion in chromosomal *aroD* gene	Invasive but nonreplicating	Yes (in monkeys)	Tests in progress
3. *Escherichia coli-Shigella flexneri* hybrids				
PGA142-1-15	*E. coli* 08 with *his* and *met* genes from *S. flexneri* 2a	Noninvasive; expresses 2a somatic antigen	None	Yes
EC104	*E. coli* K-12 with *his* and *pro* genes and 140 mDal invasion plasmid of *S. flexneri* 2a	Invasive but nonreplicating; expresses 2a somatic antigen	Yes (in monkeys)	No (insufficient attenuation)
4. *Salmonella typhi-Shigella sonnei* hybrid				
5076-1C	*S. typhi galE* mutant carrying the 120 mDal plasmid of *S. sonnei*	Invasive but non-replicating; expresses group D somatic antigen	Inconsistent	Yes

this vaccine was fed to human volunteers, and these revertants were fully virulent (DuPont et al. 1972).

Another avirulent mutant strain was obtained by serial passage on nutrient agar slants. The T32Istrati strain, denoting 32 transfers, is unable to cause keratoconjunctivitis in the guinae pig eye (Sereny test). This vaccine has been given to large groups of Romanian and Chinese children, and it has been reported to be safe and protective against both homologous and heterologous *Shigella* serotypes (Meitert et al. 1984). Unfortunately, the T32 vaccine is administered in three large doses (10^{10} CFU) annually, and this requirement limits its usefulness in most underdeveloped countries. The basis of attenuation in T32 is unclear, and it has not been tested in western populations.

The third type of avirulent mutant is the streptomycin-dependent (SmD) strains isolated by Mel et al. (1965). Since these strains cannot grow in the absence of streptomycin, they cannot colonize the intestine. It is also possible that the SmD vaccines had a substantial proportion of plasmid-defective organisms because 75% of streptomycin-independent revertants were Sereny negative (DuPont et al. 1972). SmD strains were found to be safe and effective in large scale Yugoslav field trails, but subsequent trials in the United States showed varying degrees of efficacy (reveiwed in Formal and Levine 1984). Although the SmD vaccines have provided demonstrable serotype-specific protection in human challenge studies and in field trials, they are not in current use because they require multiple priming doses and annual boosters to maintain protective immunity. In addition, the occurrence of streptomycin-independent revertants in many vaccine lots discouraged further work with SmD strains.

ATTENUATED INVASIVE HYBRID VACCINES BASED ON *S. flexneri*

In the early sixties it was found that conjugal transfer of a region located between the *rha*$^+$ and *xyl*$^+$ *loci* on the *Escherichia coli K-12* chromosome resulted in the attenuation of *S. flexneri* recipients (Falkow et al. 1963). These hybrids retain the invasive phenotype, but they do not multiply efficiently in the intestinal mucosa. *S. flexneri* carrying the *E. coli K-12* *xyl*$^+$ -*rha*$^+$ region were safe when given to monkeys in two 10^{10} CFU doses, and they elicited serotype-specific immunity against shigella challenge (Formal et al. 1966). Unfortunately, one of these stains, *S. flexneri 2a M22-18X16*, caused disease in 37% of human volunteers who ingested 10^8 CFU (DuPont et al. 1972), so it was concluded that the *xyl*$^+$ -*rha*$^+$ hybrids were not sufficiently attenuated for use as vaccines in human populations.

A current attempt to construct auxotrophic *S. flexneri* vaccines has made use of an insertionally inactivated *aroD* gene which was transduced into *S. flexneri* serotype Y strain (Lindberg et al. 1988). The resulting mutant (designated Sf1-114) retains the invasive phenotype, but a requirement for para-aminobenzoic acid limits the ability of the strain to replicate within mammalian tissues. Four doses of Sf1-114 (2-3 x 10^{10} CFU) were safe and effective in protecting monkeys against a challenge with the virulent parent strain (Lindberg et al. 1988). Human field trials with Sf1-114 are currently being conducted in Vietnam, and preliminary results indicate that doses of 10^9 CFU are well tolerated in this population (A.A. Lindberg, unpublished data).

A NONINVASIVE HYBRID VACCINE BASED ON *E. coli*

In the middle seventies, a hybrid vaccine was constructed by the conjugal transfer of the His and Met markers from *S. flexneri 2a* to a commensal *E. coli* 08 recipient. Because the *S. flexneri* group 3,4 and type II somatic antigen are linked to these markers, the resulting hybrid (PGAI42-1-15) expresses the *2a* serotype of the donor rather than the 08 serotype of the recipient (Levine et al. 1977). This hybrid was safe when administered to human volunteers in doses of 10^{10}CFU; however, three doses did not elicit demonstrable immunity against a *S. flexneri 2a* challenge (Levine et al. 1977). The failure of this non-invasive *E. coli* hybrid vaccine, along with the previous successful immunization of monkeys with the invasive *S. flexneri* hybrids (Formal et al. 1966 and Lindberg et al. 1988), suggests that invasion of the intestinal mucosa is a desirable vaccine attribute.

INVASIVE HYBRID VACCINES BASED ON *E. coli*

With the discovery of genes encoding the invasive phenotype on 120-140 mDal shigella plasmids, the potential for constructing enteroinvasive vaccines based on an avirulent *E. coli* recipient was realized. For example, mobilization of the 140 mDal plasmid from *S. flexneri* 5 into *E. coli K-12* transferred the invasive phenotype. Conjugal transfer of the His and Pro markers from the *S. flexneri 2a* chromosome into one of these invasive hybrids, resulted in a vaccine (EC104) expressing group 3,4 and type II somatic antigen (Formal et al. 1984). Other virulence- associated regions of the shigella chromosome linked to *xyl-rha* and *purE* (Sansonetti et al. 1983) were not transferred to the EC104 vaccine strain.

The *rha-mtl* region of the *S. flexneri* chromosome encodes the aerobactin iron-binding system (Griffiths et al. 1985), and insertional inactivation of the aerobactin gene in *S. flexneri 2a* (Nassif et al. 1987) results in decreased virulence somewhat similar to that observed in the M22-18X16 xyl^{+}-rha^{+} hybrid (Formal et al. 1965). The *purE* locus is linked to the *kcpA* gene which is necessary for expression of the Sereny positive phenotype in *S. flexneri* (Formal et al. 1971). Current data indicate that the $kcpA^{+}$ phenotype is associated with the ability of intracellular shigellae to translocate to adjacent cells in a HeLa cell monolayer and to form areas of cytopathic effect or "plaques" (Oaks et al. 1985). *E. coli* K-12 hybrids such as EC104, which have not acquired the $kcpA^{+}$ phenotype from *S. flexneri,* do not usually form plaques (Pal et al. manuscript in press). Along with the aerobactin-negative phenotype, this plaque-negative phenotype may be an important attenuating characteristic of the EC104 vaccine.

Since the *xyl-rha* and the *purE* genes of *S. flexneri* are not present in enteroinvasive *E. coli* K-12 hybrid vaccines, it was hoped that these strains would be both stable and non-reactive. Indeed, three doses (10^{10} CFU) of EC104 were both safe and effective in protecting monkeys against *S. flexneri 2a* challenge (Formal et al. 1984). Unfortunately, a similar *E. coli* K-12 hybrid strain elicited diarrhea in 20% of human volunteers when given at a 10^{9} dosage, and one volunteer experienced some symptoms of dysentery (Harrington and Levine, unpublished observation).

Although the pathogenic mechanism underlying these reactions is unclear, three hypotheses can be advanced. First of all, the diarrhea may be the result of the low level of *S. dysenteriae* Type 1-like enterotoxin (SLT) which is produced by *E. coli* K-12

(O'Brien et al. 1982). *S. dysenteriae* Type 1 toxin causes repetitive bursts of action potential in the intestine (Mathias et al. 1980), so SLT could also be responsible for the tenesmus experienced by some volunteers who ingested 10^9 CFU of an enteroinvasive K-12 hybrid. In addition to proteinaceous enterotoxins, a diarrheagenic effect of endotoxin delivered to the capillary bed of the lamina propria by gram-negative organisms has been proposed (Mathan et al. 1987). If invasion of the intestinal mucosa by *E. coli* elicits diarrhea even though no unique shigella virulence determinant is expressed, the use of enteroinvasive *E. coli* K-12 vaccines will obviously be limited. Finally, plaque forming variants of an *E. coli* K-12 hybrid carrying the 140 mDal plasmid from *S. flexneri* occur at a rate of approximately 10^{-6}, and this phenotype is stable upon subculture (Oaks et al. 1985; Pal et al. manuscript in press). Since such variants are similar to *kcpA*$^+$ *S. flexneri* in the plaque assay, they may also be capable of intercellular spread in the intestinal mucosa, and they may elicit concomitant symptoms of bacillary dysentery.

A HYBRID VACCINE BASED ON *SALMONELLA TYPHI* Ty21a

In addition to *E. coli* K-12, the Ty21a galactose epimeraseless (*galE*) mutant of *S. typhi* (Germanier and Fuer 1975) has been used as a recipient for shigella genes. Since the Ty21a strain is presumed to translocate across the intestinal wall only through absorptive M cells, it was deemed a promising carrier strain for delivery of heterologous somatic antigens to the gut-associated lymphoid tissue without the generalized mucosal invasion of *E. coli-S. flexneri* hybrids (Sansonetti et al. 1983). Genes encoding the *S. sonnei* O-specific side chain are carried on a 120 mDal plasmid, and mobilization of this plasmid into Ty21a resulted in a hybrid expressing the shigella group D somatic antigen (Formal et al. 1981). Like the Ty21a recipient, the hybrid strain (5076-1C) was safe for human volunteers, and three doses of 10^9 CFU provided 40% protection against *S. sonnei* challenge (Black et al. 1987).

Unfortunately, this moderate degree of protection is not consistently produced by different lots of vaccine (Black et al. 1987). It is possible that the inconsistent efficacy of this vaccine is due to the unstable association of the form I (group D) antigen with the bacterial surface. The O-disaccharide repeat units of the *S. sonnei* form I lipopolysaccharide are not covalently linked to the salmonella core lipid A in the 5076-1C hybrid (Seid et al. 1984). Therefore, the *E coli* O antigen core will be transferred into the Ty21a-*S. sonnei* hybrid so that the *S. sonnei* O-side chain can be covalently attached in the lipopolysaccharide. It is hoped that this modification will enhance the immunogenicity of the Ty21a-based vaccines. Another possibility is to use other *S. typhi* vaccine strains as antigen carriers. For example, *aroA* mutants of *S. typhi* may prove to be more stable and immunogenic than the *galE* mutant.

CONCLUSIONS

Since the ID_{50} of *S. flexneri* in humans is 10^3 CFU, invasion of the intestinal epithelium by only a few ingested organisms is apparently sufficient to cause disease. These intracellular shigellae are essentially sequestered from mucosal antibody during much of their life cycle, and they present a transient target for the humoral immune system. Therefore, an effective shigella vaccine should evoke both cellular and antibody responses. In the final analysis, it may be impossible to elicit such immune responses in the intestinal mucosa without evoking some symptoms of disease.

Nonetheless, we hope that continued basic applied research in the fields of microbial genetics and cellular immunology may yet allow the construction of safe and effective *Shigella* vaccines.

REFERENCES

Black RE, Levine MM, Clements ML, Losonsky G, Herrington D, Berman S, Formal SB (1987) Prevention of shigellosis by a *Salmonella typhi-Shigella sonnei* bivalent vaccine. J Infect Dis 155:1260-1265

Buysse JM, Stover CK, Oaks EV, Menkatesan M, Kopecko DJ (1987) Molecular cloning of invasion plasmid antigen (*ipa*) genes from *Shigella flexneri*: Analysis of *ipa* gene products and genetic mapping. J Bact 169:2561-2569

DuPont HL, Hornick RB, Dawkins AT, Snyder MJ, Formal SB (1969) The response of man to virulent *Shigella flexneri 2a*. J Infect Dis 119:296-299

DuPont HL, Hornick RB, Snyder MJ, Libonati JP, Formal SB, Gangarosa EJ (1972) Immunity in Shigellosis. II. Protection induced by oral live vaccine or primary infection. J Infect Dis 125:12-16

Dzikidze EK, Stasilevich ZK, Kavtaradze KN, Levenson VI, Subbotina YuL, Rukhadze EZ, Fedosova VG (1986) Experimental study of shigella ribosomal vaccine in monkeys. In: Ryc M, Franek J (eds) Bacteria and the Host, Czechoslovak Medical Press, Prague, p 365

Falkow S, Schneider H, Baron S, Formal SB (1963) Virulence of *Escherichia-Shigella* genetic hybrids for the guinea pig. J Bact 86:1251-1258

Formal SB, Levine MM (1984) Shigellosis. In: Germanier R (ed) Bacterial Vaccines. Academic Press, New York, p 167

Formal SB, Kent TH, May HC, Palmer A, Falkow S, LaBrec EH (1966) Protection of monkeys against experimental shigellosis with a living attenuated oral polyvalent dysentery vaccine. J Bact 92:17-22

Formal SB, Maenza RM, Austin S, LaBrec EH (1967) Failure of parenteral vaccines to protect monkeys against experimental shigellosis. Proc Soc Exp Biol Med 25:347-349

Formal SB, Gemski P, Baron LS, LaBrec EH (1971) A chromosomal locus which controls the ability of *Shigella flexneri* to evoke keratoconjunctivitis. Infec Immun 3:73-79

Formal SB, Baron LS, Kopecko DJ, Washington O, Powell C, Life CA (1981) Construction of a potential bivalent vaccine strain: Introduction of *Shigella sonnei* form I antigen genes into the *galE Salmonella typhi Ty21a* typhoid vaccine strain. Infect Immun 34:746-750

Germanier R, Fuirer E (1975) Isolation and characterization of *galE* mutant *Ty21a* of *Salmonella typhi*: A condidate strain for a live, oral, typhoid vaccine. J Infect Dis 131:553-558

Griffiths E, Stevenson P, Hale TL, Formal SB (1985) Synthesis of aerobactin and 76,000-dalton iron-regulated outer membrane protein by *Escherichia coli* K-12-*Shigella flexneri* hybrids and by enteroinvasive strains of *Escherichia coli*. Infect Immun 49:67-71

Hale TL, Oaks EV, Formal SB (1985) Identification and antigen characterization of virulence-associatd, plasmid-coded proteins of *Shigella* spp. and enteroinvasive *Escherichia coli*. Infect Immun 50:620-629

Higgins AR, Floyd TM, Kader MA (1955) Studies in shigellosis. III. A controlled evaluation of a monovalent *Shigella* vaccine in a highly endemic environment. Amer J Trop Med Hyg 4:281-288

Kopecko DJ, Holcombe J, Formal SB (1979) Molecular characterization of plasmids from virulent and spontaneously occurring avirulent colonial variants of *Shigella flexneri*. Infect Immun 24:580-582

LaBrec EH, Schneider H, Magnani TJ, Formal SB (1964) Epithelial cell penetration as an essential step in the pathogenesis of bacillary dysentery. J Bact 88:1503-1518

Levine MM, Woodward WE, Formal SB, Gemski P, DuPont HL, Hornick RB, Snyder MJ (1977) Studies with a new generation of oral attenuated shigella vaccine: *Escherichia coli* bearing surface antigens of *Shigella flexneri*. J Infect Dis 136:577-582

Lindberg AA, Karnell E, Stocker BAD, Katakura S, Sweiha H, Reinholt FP (1988) Development of an auxotropic oral live *Shigella flexneri* vaccine. Vaccine 6:146-150

Mathan VI, Mathan M, Rowley D (1987) Twenty Third Joint Conf. on Cholera, U.S. - Japan Cooperative Medical Science Program. Abstract

Mathias JR, Carlson GM, Martin JL, Shields RP, Formal SB (1980) *Shigella dysenteriae* I enterotoxin: Proposed role in pathogenesis of shigellosis. Am J Physiol 239:G382-G386

Meitert T, Pencu E, Ciudin L, Tonciu M (1984) Vaccine strain *Sh. flexneri* T32-Istrati. Studies in animals and in volunteers. Anti-dysertery immunoprophylaxis and immunotherapy by live vaccine Vadizen (*Sh. flexneri* T32-Istrati). Arch Roum Path Exp Microbiol 43:251-278

Mel DM, Terzin AL, Vuksic L (1965) Studies on vaccination against bacillary dysentery. 3. Effective oral immunization against *Shigella flexneri 2a* in a field trial. Bull WHO 32:647-655

Nassif X, Mazert MC, Mounier J, Sansonetti PJ (1987) Evaluation with an *iuc*: Tn*10* mutant of the role of aerobactin production in the virulence of *Shigella flexneri*. Infect Immun 55:1963-1969

Oaks EV, Wingfield ME, Formal SB (1985) Plaque formation by virulent *Shigella flexneri*. Infect Immun 48:124-129

O'Brien AD, LaVeck GD, Thompson MR, Formal SB (1982) Production of *Shigella dysenteriae* type 1-like cytotoxin by *Escherichia coli*. J Infect Dis 146:763-769

Sansonetti PJ, Kopecko DJ, Formal SB (1982) Involvement of a plasmid in the invasive ability of *Shigella flexneri*. Infect Immun 35:852-860

Sansonetti PJ, Hale TL, Dammin GJ, Kapfer C, Collins HH, Formal SB (1983) Alterations in the pathogenicity of *Escherichia coli* K-12 after transfer of plasmid and chromosomal genes from *Shigella flexneri*. Infect Immun 39:1392-1402

Seid RC, Kopecko DS, Sadoff JC, Schneider H, Baron LS, Formal SB (1984) Usual lipopolysaccharide antigens of *Salmonella typhi* oral vaccine strain expression the *Shigella sonnei* form I antigen. J Biol Chem 259:9028-9034

The Enteric Immune Response to Shigella Antigens

D.F. Keren, R.A. McDonald, J.S. Wassef, L.R. Armstrong, and J.E. Brown

The Department of Pathology, University of Michigan Medical School, Ann Arbor, Michigan 48109, USA

INTRODUCTION

The optimal regimen for stimulating a secretory IgA response in the intestine has yet to be determined. It is well known that parenteral administration of antigen will result in the formation of a systemic immune response directed to specific determinants on that antigen. Depending on the characteristics of the antigen, its dose, and the genetic capabilities of the animal, a humoral and/or cellular immune responses will result. While similar mechanisms must occur to stimulate immunity to antigens which are present in mucosal surfaces, including the gastrointestinal tract, less is known about the specific form or dose of antigens which would best elicit the production of immunity in the mucosa itself. It has been known for over a century that oral administration of antigens can elicit protection to some enteric infections. The discovery that IgA is the main antibody on mucosal surfaces provided the key for beginning definitive work to understand the biology of the mucosal immune system (Tomasi et al. 1965). While many tissues (bronchial mucosa, mammary glands, conjunctiva, genitourinary tract, biliary tract, etc.) are involved with the mucosal immune response, the gastrointestinal tract is overwhelmingly the major site of antigenic stimulation and immune response for secretory IgA (Brandtzaeg 1985). Recent studies indicate that a combination of parenteral and oral administration of antigens may enhance the initial secretory IgA response to *Shigella flexneri* (Keren et al. 1988). This review explores the mechanism for stimulation of the secretory IgA memory response to shigella lipopolysaccharide (LPS) and to Shiga toxin. Further, evidence is presented to implicate M cells in the initial mucosal ulcerations seen in dysentery.

CHRONIC INTESTINAL LOOP MODEL FOR STUDYING MUCOSAL IMMUNITY

Our Laboratory has been studying several aspects of the secretory IgA response to enteropathogens by using a chronically isolated ileal loop model in rabbits. We use this model as a probe to follow secretory IgA responses in rabbits given antigen by various routes (Keren et al. 1975). For this model, 3 kg New Zealand white rabbits (specific pathogen-free) are anesthetized with xylazine and ketamine. A midline abdominal incision is made and the terminal ileum is identified. 20 cm of ileum containing a Peyer's patch is isolated with its vascular supply intact. Silastic tubing is sewn into each end of the isolated segment. The free ends of the tubing are brought out through the midline incision and are tunneled subcutaneously to the nape of the neck where they are exteriorized and secured. Intestinal continuity is restored by an end-to-end anastomosis. The abdominal incision is closed in two layers. Each day, the secretions (2-4 ml) that collect in the ileal loop can be expelled by injecting air into one of the silastic tubings. Mucus is separated by centrifugation. The slightly opaque, colorless supernatant is available to study of specific immunoglobulin content and activity.

Current Topics in Microbiology and Immunology, Vol. 146

This chronically isolated ileal (Thiry-Vella) model system has been used to study the intestinal IgA response to cholera toxin, *Shigella flexneri*, *Salmonella typhi*, and Shiga toxin (Keren 1978, 1982; Yardley et al. 1978). This model system has established that multiple oral immunizations with live *Shigella flexneri* antigens are superior to parenteral immunization in eliciting a secretory IgA response (Keren et al. 1982b). However, by priming animals with a single parenteral dose of heat-killed shigella one day prior to oral challenge, the initial (primary) mucosal immune response can be improved (Keren et al. 1988). Since only secretions were tested in these studies, it was not completely certain that the gut-associated lymphoid tissues (GALT) were responsible for the IgA activity demonstrated. Therefore, our recent studies have examined the IgA production by GALT cells from these animals.

LOCATION OF IgA PRECURSOR B LYMPHOCYTES FOR *SHIGELLA FLEXNERI* FOLLOWING PRIMING FOR A MUCOSAL MEMORY RESPONSE

We originally predicted that only strains of shigella which were invasive would be effective immunogens when given orally for stimulating the mucosal memory response in intestinal secretions. To test this idea, we examined four strains of shigella with different invasive capabilities (Table 1). *Shigella flexneri* M4243 contains the 140 megadalton virulence plasmid, gives a positive Sereny test, and invades the intestinal epithelium. *Shigella* X16 contains the virulence plasmid and can invade the intestinal epithelium, however, it does not replicate following this invasion and does not give a positive Sereny test. *S. flexneri* 2457-0 contains the virulence plasmid, but does not invade the surface epithelium and does not give a positive Sereny test. The last strain studied, *S. flexneri* M4243A1 lacks the virulence plasmid and does not invade the intestinal epithelium. Surprisingly, when administered orally, all four strains were able to elicit vigorous mucosal memory responses (Keren et al. 1985, 1986). Indeed, the strongest response was elicited by the avirulent M4243A1 strain (Keren et al. 1986).

Table 1. Characteristics of Shigella used in the present studies

Strain	Virulence plasmid	Sereny test	Intestinal invasion
S. flexneri M4243	+	+	+
Shigella X16	+	o	+
S. flexneri 2457-0	+	o	o
S. flexneri M4243A1	o	o	o

Since all four strains could elicit a mucosal memory response, we wished to determine the location of specific antigen-reactive cells following different priming regimens and to establish their migration pattern after oral challenge. In these studies, we used *S. flexneri* M4243A1 for immunization. Three intragastric immunizations were given one week apart at 74, 67, and 60 days prior to dissection. One to ten days before dissection, the rabbits were given a single oral dose of live *S. flexneri* M4243A1. The rabbits were sacrificed and the lymphoid populations in the spleen, Peyer's patches, and mesenteric lymph nodes were sampled. Tissues were cut into 1 cm^3 fragments with a sterile blade and placed in cold RPMI-1640. The cells were carefully teased apart and passed through the steel mesh. This material was centrifuged at 400 x *g* at room temperature for 7 minutes. The pellet was gently resuspended and washed twice in RPMI 1640 medium. The total number of cells and their viability were determined. 4 x 10^6 mononuclear cells in 1 ml tissue culture medium were added to each well of 24 well polystyrene tissue culture plates with flat bottom wells (Costar). Cultures were placed in a humidified, 5% CO_2, 37°C incubator. At the times indicated (days in culture), three wells for each tissue were aspirated. Cellular debris was removed by centrifugation at 400 x *g* for 5 minutes and the supernatants were stored at -20°C until they were assayed. Assays were performed using a previously described ELISA specific for IgA and IgG antibodies to *S. flexneri* LPS (Keren 1979).

In Tables 2, 3, and 4 are shown the IgA anti-Shigella LPS responses from rabbit Peyer's patches, mesenteric lymph node, and spleen, respectively. It is clear from these data that as early as one day following rechallenge with live $M4243A_1$, IgA-specific B lymphocytes are present within Peyer's patches. By the fourth day following rechallenge, a significant increase is seen in the amount of antigen-specific IgA produced. Strikingly, by the fifth day, B cells have left the Peyer's patches only to return by the tenth day following rechallenge. Indeed, by day ten, considerable antigen-specific IgA can be detected as soon as the second day in tissue culture. This implies that the cells present in Peyer's patches were present in considerably greater numbers at the later times.

In contrast, the lymphocytes from mesenteric lymph nodes from the first day following rechallenge were unable to produce specific IgA anti-Shigella LPS. By the third postchallenge day, there was a dramatic increase in the IgA anti-Shigella LPS which persisted through the fourth day in mesenteric lymph node cells. However, by the fifth day postchallenge these responses had returned to baseline values indicating that the cells traveled from the Peyer's patches to the mesenteric lymph nodes within three days following oral rechallenge. They left this station such that by day five postchallenge little specific activity to Shigella LPS was detectable. It is likely that these cells then traveled to the spleen as cultures from splenic mononuclear cells on days one and three show virtually no IgA anti-Shigella LPS activity, while the cultures from day four show significant IgA anti-Shigella LPS activity (Table 4). The period of time within the spleen is very brief as the activity declined by day five to almost baseline values. Beyond day six, no significant IgA anti-Shigella LPS activity was detectable.

When these same tissues were examined in an unprimed rabbit, no significant IgA anti-Shigella LPS activity was detectable from any tissue on any day. Studies of the IgG content of these supernatants have been most instructive. Consistent with our former *in vivo* data indicating that little IgG anti-Shigella LPS is present in intestinal secretions (Keren et al. 1978; Keren et al. 1986), our present studies found little specific IgG produced in the culture supernatants from the mononuclear cell preparations. This indicates that the responses seen in the secretions do not merely reflect preferential uptake of systemic polymeric IgA by intestinal epithelial cells with subsequent transport into the gut lumen. Rather, they accurately reflect the capabilities of the mononuclear cells stimulated following oral antigen administration to produce an IgA as opposed to an IgG response to *Shigella flexneri* antigens.

Table 2. IgA anti-Shigella LPS in Peyer's patch supernatant

Days in culture	Day after challenge 1	3	4	5	6	10
0	0.008*	0.009	0.000	0.003	0.014	0.002
1	0.016	0.035	0.078	0.017	0.047	0.003
2	0.034	0.012	0.083	0.009	0.061	0.963
3	0.120	0.019	0.173	0.004	0.001	1.963
4	0.119	0.024	0.367	0.063	0.351	1.255
5	0.077	0.059	0.479	0.128	0.440	2.112
6	0.186	0.146	0.503	0.085		2.005
7	0.245	0.000	0.667	0.009	0.093	1.797
14	0.241	0.026	1.284	0.055	0.424	1.371
21	0.438	0.332	1.871	0.063	0.815	1.107

*Results expressed as ΔOD 405nm/200 min

Table 3. IgA anti-Shigella LPS in mesenteric lymph node supernatant

Days in culture	Day after challenge 1	3	4	5	6	10
0	0.000	0.051	0.000	0.008	0.012	0.004
1	0.006	0.544	0.644	0.008	0.049	0.002
2	0.009	0.500	0.326	0.001	0.036	0.009
3	0.002	1.779	0.375	0.027		0.008
4	0.006	0.664	0.433	0.008	0.097	0.007
5	0.022	1.693	0.544	0.027	0.000	0.006
6	0.008	1.061	0.464	0.002		0.009
7		0.434	0.377	0.001	0.042	0.010
14	0.030	0.910	0.727	0.026	0.082	0.021
21	0.016	2.273	0.413	0.021	0.033	0.005

Table 4. IgA anti-Shigella LPS in spleen supernatant

Days in culture	Day after challenge 1	3	4	5	6	10
0	0.000	0.053	0.000	0.008	0.000	0.002
1	0.000	0.022	0.271	0.013	0.001	0.000
2	0.006	0.015	0.364	0.000	0.000	0.003
3	0.000	0.025	0.372	0.021	0.006	0.010
4		0.032	0.673	0.013	0.039	0.014
5	0.001	0.088	0.740	0.048	0.010	0.026
6	0.014	0.107		0.176		0.012
7	0.010			0.116	0.007	0.038
14	0.040	0.050	1.134	0.076		0.019
21	0.034	0.056	0.878	0.058	0.007	0.034

M CELL UPTAKE OF SHIGELLA AND THE MUCOSAL IMMUNE RESPONSE

Since our previous studies demonstrated that oral or intraloop immunizations of all four strains of *S. flexneri* (Table 1) were able to elicit a vigorous primary mucosal immune response and a mucosal memory response, we hypothesized that these strains must be processed in a similar manner by the GALT. We know from the work of Owen (1977) that there are specialized surface epithelial cells which sample intraluminal antigens including microorganisms (Bockman and Cooper 1973; Rosner and Keren 1984). It is clear that M cells respond to microorganisms present within the gut lumen. A recent study by Smith et al. (1987) demonstrated that a significant maturation and increase in the number of M cells occurs when specific pathogen-free mice were transferred to a normal animal house environment.

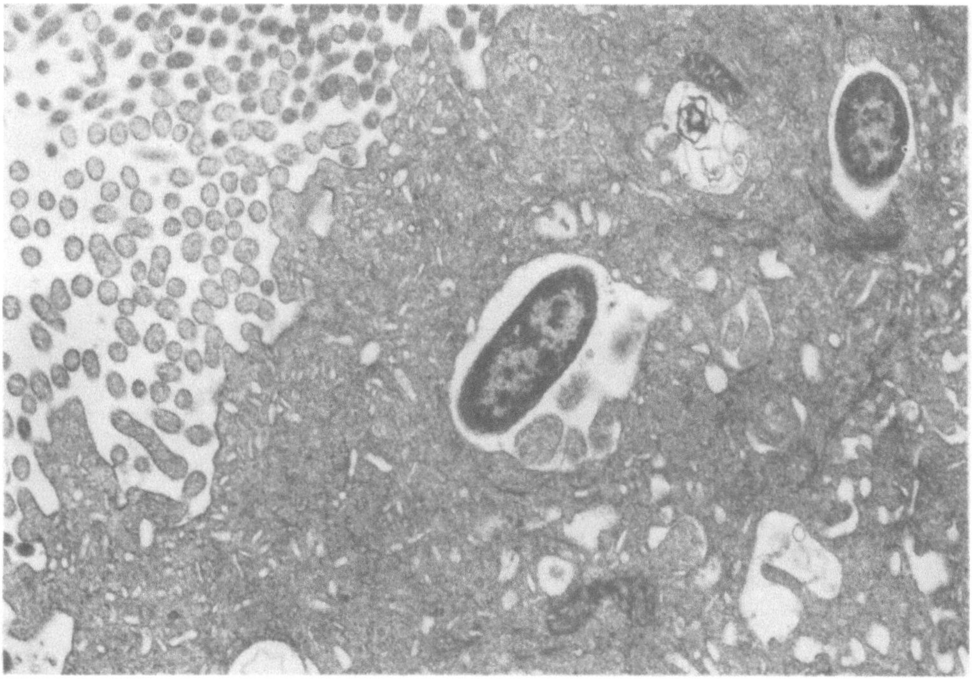

Fig. 1. Electron photomicrograph showing two Shigellae within an M cell

To determine how *S. flexneri* are processed by the intestinal epithelium, we allowed the four strains of Shigella listed in Table 1 to incubate for 90 minutes or 18 hours in acutely ligated loops of rabbit ileum. All four strains of Shigella showed readily demonstrable uptake over the dome regions of the Peyer's patches. This was assessed in two ways. First, ultrastructural studies were performed to demonstrate the bacteria within the M cells in the follicle-associated epithelium (Fig. 1). Then, frozen sections of rabbit ileum were obtained through Peyer's patches and adjacent villus epithelium. These sections were stained with Giemsa stain to demonstrate the Shigella (Fig. 2). The frozen sections allowed us to perform an accurate count of the bacteria. These counts were performed with the aid of the Bioquant Biometrics Image Analyzer (Nashville, Tennessee) with an IBM computer used to measure the actual length in millimeters of the lining epithelium over the villi and over the dome regions of the Peyer's patches. The average of 100 areas for dome and villus areas from representative rabbits was calculated. This allowed us to directly express data as bacteria/mm^2 of surface epithelium. Further, it permitted a direct comparison of villus surface area to follicle-associated epithelium surface area. To be included in a count, we required that the entire Shigella be located within the cytoplasm. By focusing up and down, a vacuole was usually discernable around each engulfed bacterium (Fig. 2). Bacteria which were adherent to the surface epithelium but which were not clearly present within the cytoplasm were not counted.

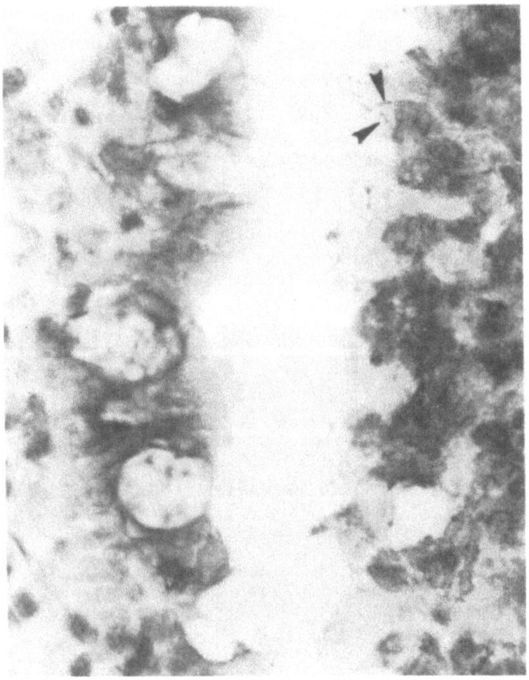

Fig. 2. Photomicrograph depicting Shigellae (arrows) within the surface epithelium

The bacteria seen at the 90 minute time period by ultrastructural studies were contained within membrane lined vesicles (Fig. 1), although some vesicles in the loops given the pathogenic strain (*S. flexneri* M4243) showed early evidence of breakdown of the vesicles. The three nonpathogenic strains (Sereny test-negative) were taken up with equal efficiency regardless of their invasive capabilities or of the presence of the 140 megadalton virulence plasmid. All strains examined had relatively few Shigella (ten fold less) within the villus epithelium as compared to the follicle-associated epithelium. The pathogenic strain *S. flexneri* M4243 had significantly greater uptake of the bacteria in the dome regions than did the three nonpathogenic strains (Fig. 3).

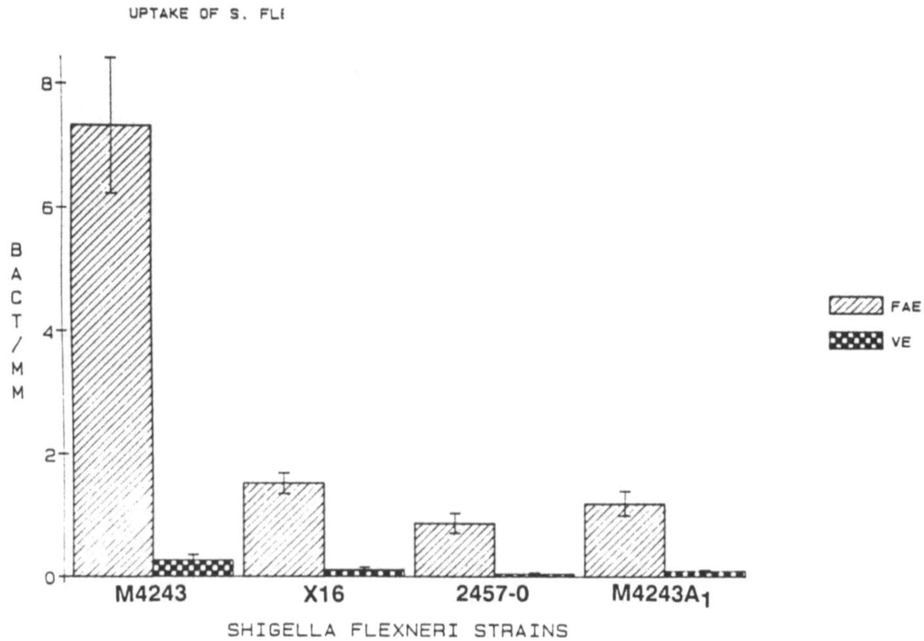

Fig. 3. Uptake of Shigellae by follicle-associated epithelium (FAE) and villus epithelium (VE)

Both pathogenic and nonpathogenic strains of Shigella were taken up preferentially by the specialized M cells in the follicle-associated epithelium as opposed to the villus epithelium; therefore, M cells do not distinguish between Shigella on the basis of expression of antigens encoded by the virulence plasmid. Since all four strains have been found to elicit significant mucosal immune responses in our previous studies where direct intraintestinal stimulation was given to chronically isolated ileal loops and since the three non-invasive strains could prime rabbits for mucosal memory response regardless of their ability to invade surface epithelium or the presence of the 140 megadalton virulence plasmid, we believed that the strains would be sampled with equal efficiency by the surface M cells. The findings in these acute loop studies are consistent with this hypothesis. There was, however, a significant difference of the uptake of the pathogenic *S. flexneri* M4243 strain versus the avirulent strains at 90 minutes. This likely reflected the successful replication by the latter bacteria within the tissue following uptake. Therefore, we followed this process for 18 hours to allow replication to continue and pathologic events to occur. After 18 hours of incubation, profound mucosal ulceration was seen exclusively with the *S. flexneri* M4243 strain. The acute loops incubated for 18 hours with these bacteria showed a hemorrhagic surface with marked acute inflammation throughout the lamina propria. Ulceration was present predominantly in the dome regions over the Peyer's patches. Although there was mucosal damage in the adjacent villi, the surface epithelium was, in general, intact. With the pathogenic M4243 strain, myriads of microorganisms were seen in the exudate over the ulcer and within the tissues, attesting to their successful replication.

In marked contrast, Shigellae were not found within the surface epithelium of acute loops incubated with the noninvasive strains for this time. Further, the three nonpathogenic strains showed no ulceration after the 18 hour incubation. With the *Shigella* X16 strain, there was some hemorrhage in the lumen. However, the epithelium overlying the villi and the dome regions of Peyer's patches in animals given the *Shigella* X16 strain was intact.

These findings indicate that in addition to being the site for antigen sampling , M cells serve as the portal of entry for pathogenic microorganisms. Indeed, M cells have been proposed by others as a portal of entry for intestinal pathogens including the human immunodeficiency virus (Sneller and Strober 1986).

MUCOSAL IMMUNE RESPONSE TO SHIGA TOXIN: FUNCTIONAL SIGNIFICANCE

While invasion with replication in the mucosa is the main mechanism in the pathogenesis of dysentery, it has been known for some time that *Shigella dysenteriae* 1 produces a protein cytotoxin, Shiga toxin, which in tissue culture inhibits protein synthesis in specific cell lines (Sandvig 1987; Brown 1980; Eiklid and Olsnes 1980). Since Shiga toxin must first bind to a glycolipid receptor before it can initiate cell change *in vitro* (Brown et al. 1983), the presence of an antibody to interfere with this binding would theoretically interfere with the cytotoxicity process.

Vigorous mucosal immune responses can be elicited to other enterotoxins (Yardley et al. 1978; Pierce et al. 1983). Studies using the present Thiry-Vella loop model system have stressed that cholera toxin is a most potent mucosal immunogen (Yardley et al. 1978). In the present studies, we immunized two rabbits intraloop with a preparation of Shiga toxin provided by Dr. J. Edward Brown. The results in Fig. 4 show the specific intestinal antibody activity from two rabbits given direct intraloop immunization with Shiga toxin. The Shiga toxin was given on days 0, 7, and 14. We found that by day 10, a significant increase in the IgA anti-Shiga toxin activity over background had occurred. Booster doses of Shiga toxin given on days 7 and 14 had the effect of decreasing both the IgA anti-Shiga toxin activity in the following day's secretion as well as decreasing the inhibition titer of the loop secretions performed in the HeLa cell assay. This was due to binding of specific IgA in these secretions by the Shiga toxin. Both doses were followed by substantial increased IgA activity in subsequent days. There was an excellent correlation of the IgA anti-Shiga toxin activity with the inhibition titer. Little IgG anti-Shiga toxin was detected in these intestinal secretions. Future studies will attempt to confirm these results in a larger group of animals and to establish whether a secretory IgA memory response against Shiga toxin can be elicited. It is notable, however, that Shiga toxin is second only to cholera toxin in being able to elicit a vigorous secretory IgA response to our model system. Since it is clear that IgA responses are highly dependent on T cell control mechanisms (Kawanishi et al. 1983; Campbell and Vose 1985), it may be worthwhile to study the mechanism of Shiga toxin mucosal stimulation to determine how it interacts with helper or switch T cells.

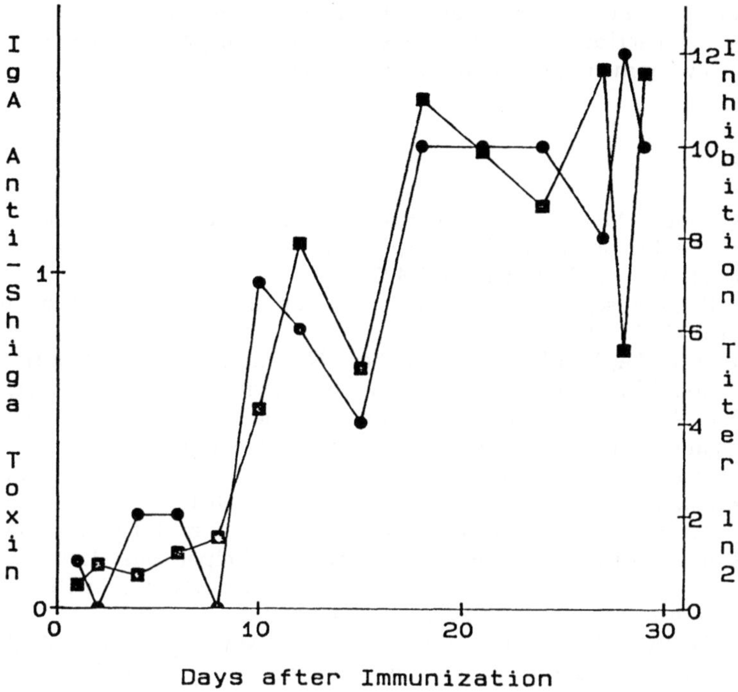

Fig. 4. Mean IgA anti-Shiga toxin activity (ELISA) (squares) correlates well with the mean cytotoxicity inhibition titer (circles) in secretions from immunized rabbits

SUMMARY

Mucosal immunity to some enteropathogens occurs naturally following infection. By learning how to optimize initiation of the mucosal immune response it will be possible to develop vaccines against a wide variety of enteropathogens and their toxic products. In the past few years, we have examined stimulation of the mucosal response to Shigella antigens. We have found that the mucosal memory response to Shigella LPS can be stimulated by oral immunization with live, but not with killed Shigella. This primes specific B lymphocytes which, following rechallenge, quickly migrate from the Peyer's patches to mesenteric lymph nodes, the spleen, and back to the Peyer's patches. We have found that the uptake of *S. flexneri* is the initial step in developing a mucosal immune response to Shigella. Whereas there is little difference between the initial uptake of virulent and avirulent bacteria by M cells, pathogenic strains of Shigella are able to replicate following their uptake by the specialized M cells located in the follicle-associated epithelium of the gut. This likely serves as the source of the ulcerative lesions found in dysentery. Lastly, we have detected a vigorous secretory IgA response to Shiga toxin. The titer of IgA activity to Shiga toxin from these loop secretions correlated well with the ability to prevent Shiga toxin cytotoxin effects *in vitro*. The extremely vigorous mucosal immune response to Shiga toxin makes this an attractive alternative to cholera toxin to potentiate the secretory IgA immune response.

Acknowledgment. The authors thank Ms. Teri Throne for her excellent help in preparing this manuscript. This work was supported in part by U.S. Army Medical Research and Development command Contract DAMD17-85-C-5006.

REFERENCES

Bockman DE, Cooper MD (1973) Pinocytosis by epithelium associated with lymphoid follicles in the Bursa of Fabricius, appendix and Peyer's patches. An electron microscopic study. Am J Anat 136:455-477

Brandtzaeg P (1985) Research in Gastrointestinal Immunology: State of the art. Scand J Gastroenterol 114:137-156

Brown JE, Karlsson KA, Lindberg A, Stromberg N, Thurin J (1983) Identification of the receptor glycolipid for the toxin of Shigella dysenteriae. In: Chester MA, Heinegard D, Lundblad A, Svansson S (ed) Proceedings of the 7th International Symposium of Glycoconjugates, Lund, Sweden, Rahms, Lund, Sweden, p 678

Brown JE, Rothman SW, Docto BP (1980) Inhibition of protein synthesis in intact HeLa cells by *Shigella dysenteriae* 1 toxin. Infec Immun 29:98-107

Campbell D, Vose BM (1985) T-cell control of IgA production. I. Distribution, activation conditions and culture of isotype-specific regulatory helper cells. Immunol 56:81-92

Eiklid K, Olsnes S (1980) Interaction of *Shigella shigae* cytotoxin with receptors on sensitive and insensitive cells. J Recept Res 1:199-213

Kawanishi H, Salzman, LE, Strober W (1983) Mechanisms regulating IgA class-specific immunoglobulin production in murine gut-associated lymphoid tissues. I. T-cells derived from Peyer's patches which switch sIgM B cells *in vitro*. J Exp Med 157:433-449

Keren DF (1979) Enzyme-linked immunosorbent assay for IgA and IgG antibodies to *S. flexneri* antigens. Infect Immun 24:441-448

Keren DF, Collins HH, Baron LS, Kopecko DJ, Formal SB (1982A) Mucosal (IgA) immune responses of the intestine to a potential vaccine strain: *Salmonella typhi* gal E Ty 21a expressing *Shigella sonnei* Form I antigen. Infect Immun 37:387-390

Keren DF, Elliott HL, Brown GD, Yardley JLH (1975) Atrophy of villi with hypertrophy and hyperplasia of Paneth cells in isolated (Thiry-Vella) ileal loops in rabbits. Gastroenterol 68:83-93

Keren DF, Holt PS, Collins HH, Gemski P, Formal SB (1978) The role of Peyer's patches in the local immune response of rabbit ileum to live bacteria. J Immunol 120:1892-1898

Keren DF, Kern SE, Bauer DH, Scott PJ (1982B) Direct demonstration in intestinal secretions of an IgA memory response to orally administered *Shigella flexneri* antigens. J Immunol 128:475-479

Keren DF, McDonald RA, Formal SB (1986) Secretory immunoglobulin. A response following peroral priming and challenge with *Shigella flexneri* lacking the 140-megadalton virulence plasmid. Infect Immun 54:920-923

Keren DF, McDonald RA, Scott P, Rosner AM, Strubel E (1985) Effect of antigen form on local immunoglobulin A memory response of intestinal secretions to *Shigella flexneri*. Infect Immun 47:123-129

Keren DF, McDonald RA, Carey JL (1988) Combined parenteral and oral immunization results in an enhance mucosal immunoglobulin A response to *Shigella flexneri*. Infect Immun 56:910-915

Owen RL (1977) Sequential uptake of horseradish peroxidase by lymphoid follicle epithelium of Peyer's patches in the normal and unobstructed mouse intestine: An ultrastructural study. Gastroenterol 72:440-451

Pierce NF, Cray Jr WC, Saccr Jr JB, Craig JP, Germanier R, Furer E (1983) Procholeragenoid: A safe and effective antigen for oral immunization against experimental cholera. Infect Immun 40:1112-1118

Rosner AJ, Keren DF (1984) Demonstration of "M" cells in the specialized follicle-associated epithelium overlying isolated follicles in the gut. J Leukocyte Biol 35:397-404

Sandvig K, Brown JE (1987) Ionic requirements for entry of Shiga toxin from *Shigella dysenteriae* 1 into cells. Infect Immun 55:298-303

Smith MW, James PS, Tivey DR (1987) M cell numbers increase after transfer of SPF mice to a normal animal house environment. Am J Pathol 128:385-389

Sneller MC, Strober W (1986) M cells and host defense. J Infect Dis 154:737-741

Immune Response to Oral Salmonella Vaccines

A. Tagliabue

Sclavo Research Center, Via Fiorentina 1, 53100 Siena, Italy

INTRODUCTION

Infections by *Salmonella* species still represent a public health problem throughout the world. Typhoid fever, the disease induced by *S. typhi*, remains one of the most common infections in developing countries, which may lead to death, particularly if associated with malnutrition. An incidence has been reported of 540 typhoid cases/100,000 (0.5%) inhabitants in the developing countries, with a total annual estimate of 12.5 million cases in the world (excluding China) (Edelman and Levine 1986). On the other hand, salmonellosis outbreaks caused by *S. typhimurium*, *S. enteritidis*, and other species are often observed in developed countries, sometimes as a consequence of modern habits in alimentation (kitchens catering for large numbers of people, microwave ovens, etc.). Thus, an effective and safe vaccine against *Salmonellae* is still considered a high priority by WHO in its Vaccine Development Programme.

Indeed, parenteral typhoid vaccines were among the first to be used for large-scale vaccination. They usually consisted of heat-phenolized and acetone-dried killed whole *S. typhi* alone or in combination with *S. paratyphi* A and B (TAB). The efficacy of these vaccines was shown to be significant in several field trials in areas endemic for typhoid. However, the frequency and severity of side effects induced by these parenteral vaccines has rendered them unsatisfactory. In fact, at least 25% of the recipients develop fever, malaise, and notable local reactions. Therefore, new vaccines against *S. typhi* have been investigated. Since the first encounter between the host immune system and *Salmonellae* occurs at the mucosal level in the gastrointestinal tract, it appeared logical to try to develop oral vaccines directed to stimulate the mucosal responses. Indeed, Germanier and coworkers have succeeded in developing an oral vaccine against typhoid fever (Germanier and Furer 1975) which represents the first oral vaccine against bacteria, and the second since the introduction of Sabin's anti-polio oral vaccine, to be in general use. Thus, the possibility of orally immunizing against typhoid fever prompted us and other groups to investigate *in vitro* procedures able to assess vaccine-induced local immune responses and their predictive value for protection against infection. Results obtained so far are summarized and discussed below.

VACCINES AGAINST TYPHOID FEVER

Parenteral typhoid vaccines consisting of killed bacteria have been in use for decades, and their effectiveness has been demonstrated as ranging from moderate to excellent in placebo-controlled field trials in endemic areas (Table 1). However, these vaccines are unsatisfactory due to the frequency and severity of the side effects. In the last decade, new typhoid vaccines have reached field trials. The first and most widely employed so far is the so-called Ty21a oral live vaccine developed by Germanier and coworkers. This mutant, lacking the enzyme UDP galactose-4-epimerase of the Leloir pathway, was

Current Topics in Microbiology and Immunology, Vol. 146

obtained by nitrosoguanidine treatment, which also gave rise to other genetic lesions that confer to this *S. typhi* strain its unique protective characteristics. The first controlled field trial in Egypt was promising in terms of protection; however, a larger study in Chile with a new vaccine preparation in capsules gave a lower protection rate. This provided the opportunity to other groups involved in studies of oral vaccines to suggest that more effective vaccines can be obtained. At this point, two main directions were chosen: one in favor of more purified, less toxic parenteral vaccines such as the Vi antigens reported recently (Klugman et al. 1987), the other in favor of oral vaccines expected to immunize more naturally in terms of boosting mucosal immune responses (See Table 1 for summary). The final solution to this dichotomy will obviously come from field trials, safety studies and, lastly, by the capacity of these vaccines to avoid the establishment of healthy carriers of this infection. This will require long and expensive studies. In the meantime, it is likely that experiments in volunteers which are directed to measure vaccine-induced immune responses against *S. typhi* will provide useful information.

IMMUNE RESPONSES TO TYPHOID VACCINES

The first important issue to be considered in analyzing vaccine-induced immune responses to *S. typhi* antigens is the level of natural immunity of donors from areas which are, or are not endemic for typhoid fever. A humoral response such as serum antibody levels against *S. typhi*-LPS has often been tested in a variety of studies. This has been investigated chiefly in longitudinal studies, i.e. prior to and after infection or vaccination. Only recently have data become available that allow direct comparison between healthy residents in typhoid-endemic (Santiago, Chile) and typhoid-free (Maryland, USA) regions (Murphy et al. 1987). Interestingly, no differences were found in serum antibodies against O-polysaccharide from *S. typhi* between these two groups. Similarly, we found no differences in donors from an Italian endemic area (Bari) and those from a nonendemic area (Siena) when *S. typhi*-LPS was used as antigen in ELISA (Jirillo et al. submitted). In the past, seroconversion after TAB parenteral vaccine was observed in individuals from an endemic area (India) (Kumar et al. 1974). In contrast, seroconversion to *S. typhi* antigens was clearly found after live oral Ty21a vaccine only in donors from non-endemic areas, whereas vaccines from endemic areas showed a much less clear pattern of seroconversion (Murphy et al. 1987). Thus, the employment of the antibody titer to *S. typhi* antigen as a predictive parameter for vaccine effectiveness may present some difficulties.

Cell-mediated responses to *S. typhi* were also assessed mainly in donors from endemic areas before and after parenteral vaccines. Employing leukocyte migration-inhibition, Kumar et al. (1974) did not find a positive correlation with vaccination. Similarly, lymphocyte proliferation in response to *S. typhi* antigens in individuals from endemic areas was elevated before and after an oral vaccine, indicating that this parameter can only be used in an evaluation prior to contact with *S. typhi* in individuals with no history of typhoid fever (Murphy et al. 1987).

In the course of studies directed to elucidate mucosal responses to gram-negative enteropathogenic bacteria, we found that a third possibility may be important, i.e. antibody-dependent cellular cytotoxicity (ADCC). We found that lymphocytes from the gut-associated lymphoid tissues as well as from peripheral organs can express *in vitro* antibacterial activity with an ADCC-like mechanism (Nencioni et al. 1983; Tagliabue et al. 1983, 1984). In this case, a subset of T-lymphocytes with helper phenotype were

Table 1. Vaccines against *Salmonella typhi*

Name vaccine	Characteristics	Route	Country	Incidence of typhoid x 10^5	Date	Number of vaccinees	Age	Number of doses	Surveillance duration	Efficacy	Reference
L	Heat-phenol inactivated	Parenteral	Yugoslavia	1488	1960-63	5,028	2-50 yr.	2	2 1/2 yr.	51%	Yugoslav Typhoid Commission (1964)
L	Heat-phenol inactivated	Parenteral	Guyana	605	1960-67	27,756	5-15 yr.	2	7 yr.	67%	Ashcroft et al. (1967)
L	Heat-phenol inactivated	Parenteral	USSR	162	1962-65	36,999	7-15 yr.	2	2 1/2 yr.	66%	Heifec et al. (1966)
Vi	Polysaccharide	Parenteral	Nepal	1620	1986-87	3,457	5-44 yr.	1	17 m.	75%	Klugman et al. (1987)
Ty2	Acetone inactivated	Oral	India	1540	1968-69	6,060	0-17 yr.	3	15 m.	24%	Chuttani et al. (1971)
Ty21a	Live-galE mutant	Oral	Egypt	138	1978-81	16,486	6-7 yr.	3	3 yr.	96%	Wahdan et al. (1982)
Ty21a	Live-galE mutant	Oral	Chile	310	1983-86	22,170	6-21 yr.	3	3 yr.	67%	Levine et al. (1987a)
541 Ty (Vi^+)	*AroA/purA* mutant	Oral	phase 1 completed								Levine et al. (1987b)
543 Ty	*AroA/purA*	Oral	phase 1 completed								Levine et al. (1987b)
ts Ty2	Temperature sensitive mutants	Oral	- experimental stage								Morris-Hooke et al. (1986)

found to be armed by antibodies of the IgA isotype against *Salmonella* and *Shigella* species. When direct *in vitro* antibacterial activity was tested, employing peripheral blood mononuclear cells (PBMC) from normal healthy donors from an area with a low incidence of typhoid cases (4 per year/10^5 inhabitants), it was found that $CD4^+$ Leu 8^+ lymphocytes armed by IgA can kill *S. typhi* even though at high effector to target cell ratios and with low efficiency (Tagliabue et al. 1985b, see also Table 2). A schematic model of this newly described ADCC-like activity is shown in Fig. 1. Since the majority of our studies have been performed *in vitro,* the *in vivo* relevance of IgA-ADCC was tested in donors orally immunized with Ty21a. Indeed, in two independent studies with volunteers from non-endemic areas (Tagliabue et al. 1985a, 1986), it was found that oral vaccination increased IgA-ADCC expressed by peripheral CD4 lymphocytes, and that sera from vaccinees contain arming IgA. Interestingly, the group of Levine and coworkers (1987b) confirmed the reliability of the assay described by us in an independent study directed at the elucidation of the effectiveness of a new oral live vaccine with *aro A* and *pur A* mutations.

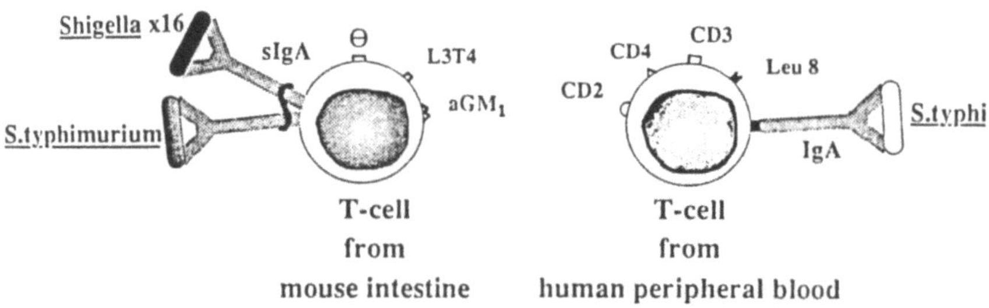

Fig. 1. Schematic representation of antibacterial IgA-ADCC

Indeed, Ty21a vaccine represents an interesting opportunity to measure the peripheral expression of mucosal immunity. In fact, employing a modification of ELISPOT described by Czerkinsky et al. (1987), it was recently shown that IgA responses in PBMC are induced by the Ty21a vaccine (Forrest 1988), thus confirming our IgA arming studies. We have also performed studies in volunteers from endemic areas. We have demonstrated in an *in vitro* assay that these healthy donors possess high antibacterial activity (Jirillo et al. submitted) (Table 2). Preliminary results would suggest that in these subjects, IgG-ADCC expressed by CD8 lymphocytes would act together with IgA-ADCC expressed by CD4 lymphocytes. No data are available as yet in vaccinated donors from endemic areas.

Table 2. *In vitro* antibacterial activity of peripheral blood lymphocytes against *S. typhi*

Experimental group	% Antibacterial activity									Mechanism
	200[a]	100	50	25	12	6	3	1.5	0.7	
Normal donors from non-endemic areas	22	14	4							IgA-ADCC by $CD4^+$ cells
Normal donors from endemic areas						47	30	27		IgA-ADCC by $CD4^+$ cells and IgG-ADCC by $CD8^+$ cells
Ty21a vaccinees from non-endemic areas	45	33	25							IgA-ADCC by $CD4^+$ cells
AIDS patients	-8	-4	12							Lack $CD4^+$ cells

[a]Effectors to bacteria ratio.

THE Tα LYMPHOCYTE

On the basis of the above mentioned results, it can be concluded that the field of typhoid vaccines is facing a renaissance, and that this renewed interest is also bringing new information to the understanding of the mucosal-induced immune response against bacteria in humans. Among other issues, it seems of interest to stress the potential importance of the $CD4^+$ $Leu8^+$ lymphocyte bearing receptors for IgA Fc, i.e. the Tα cell. Many experiments are still required in order to elucidate the relevance and effectiveness of this defense mechanism. However, a noteworthy indication comes from other *in vivo* experiments. During HIV infection, the $CD4^+$, $Leu8^+$ lymphocytes are among the preferential targets of the virus which kills them. Indeed, these patients have no IgA-ADCC against *Salmonellae* (Tagliabue et al. submitted). Even more striking, salmonellosis is now considered the most common bacterial infection associated with AIDS (Weber 1986).

REFERENCES

Ashcroft MT, Singh B, Nicholson CC, Ritchie JM, Sobryan E, Williams F (1967) A seven year field trial of two typhoid vaccines in Guyana. Lancet ii:1056-1060

Chuttani CS, Prakash K, Vergese A, Sharma U, Singha P, Ray G (1971) Effectiveness of oral killed typhoid vaccine. Bull WHO 45:445-450

Czerkinsky CC, Prince SJ, Michalek SM, Jackson S, Russell MW, Moldoveanu Z, McGhee JR, Mestecky J (1987) IgA antibody-producing cells in peripheral blood after antigen ingestion. Proc Natl Acad Sci USA 84:2449-2453

Edelman R, Levine MM (1986) Summary of an international workshop on typhoid fever. Rev Inf Dis 8:329-349

Forrest BD (1988) Identification of an intestinal immune response using peripheral blood lymphocytes. Lancet i:81-83

Germanier R, Furer E (1975) Isolation and characterization of gal E mutant Ty21a of *Salmonella typhi*: A candidate strain for a live oral typhoid vaccine. J Inf Dis 141:553-558

Heifec LB, Salmin LV, Lejtman MZ, Kuz'minova ML, Vasil'eva AV, Levina L, Bencianova TG, Pavlova EA, Antonova AA (1966) A controlled field trial and laboratory study of five typhoid vaccines in the USSR. Bull WHO 34:321-339

Jirillo E, Caretto G, Grassi PP, Nencioni L, Tagliabue A (1988) Antibody-driven T cell-mediated activity against enteropathogenic bacteria: A comparison between human volunteers from endemic and non-endemic areas for typhoid fever. (Submitted for publication)

Klugman KP, Gilbertson IT, Koornhof HJ, Schneerson R, Cadoz M, Gilbertson IT, Robbins JB, Schultz D, Armand J, Vaccination Advisory Committee (1987) Protective activity of Vi capsular polysaccharide vaccine against typhoid fever. Lancet i:1165-1169

Kumar R, Malaviya AN, Murthy RGS, Venkataraman M, Mohapatra LN (1974) Immunological study of typhoid: Immunoglobulins, C_3, antibodies, and leukocyte migration-inhibition in patients with typhoid fever and TAB-vaccinated individuals. Infect Immun 10:1219-1225

Levine MM, Ferreccio C, Black RE, Germanier R, Chilean Typhoid Committee (1987a) Large-scale field trial of Ty21a live oral typhoid vaccine in enteric-coated capsule formation. Lancet i:1049-1052

Levine MM, Herrington D, Murphy JR, Morris JG, Losonsky G, Tall B, Lindberg AA, Svenson S, Bayar S, Edwards MF, Stocken B (1987b) Safety, infectivity, immunogenicity and *in vivo* stability of two attenuated auxotrophic mutant strains of *Salmonella typhi*, 541 Ty and 543 Ty, as live oral vaccines in humans. J Clin Invest 79:888-902

Morris-Hooke A, Wang Z, Cerquetti MC, Bellanti JA (1986) Temperature-sensitive vaccine strains of *Salmonella typhi*. In: Tagliabue A, Rappuoli R, Piazzi SE (eds) Bacterial Vaccines and Local Immunity. Ann Sclavo 1-2, p 23

Murphy JR, Bagar S, Munoz C, Schlesinger L, Ferreccio C, Lindberg AA, Svenson S, Losonsky G, Koster F, Levine MM (1987) Characteristics of humoral and cellular immunity to *Salmonella typhi* in residents of typhoid-endemic and typhoid-free regions. J Inf Dis 156:1005-1009

Nencioni L, Villa L, Boraschi D, Berti B, Tagliabue A (1983) Natural and antibody-dependent cell-mediated activity against *Salmonella typhimurium* by peripheral and intestinal lymphoid cells in mice. J Immunol 130:903-907

Tagliabue A, Boraschi D, Villa L, Keren DF, Lowell GH, Rappuoli R, Nencioni L (1984) IgA-dependent cell-mediated acitivity against enteropathogenic bacteria: Distribution, specificity and characterization of the effector cells. J Immunol 133:988-992

Tagliabue A, Nencioni L, Caffarena A, Villa L, Boraschi D, Cazzola G, Cavalieri S (1985a) Cellular immunity against *Salmonella typhi* after live oral vaccine. Clin Exp Immunol 62:242-247

Tagliabue A, Nencioni L, Mantovani M, Lazzarin A, Villa L, Romano M, Rossini S, Uberti-Foppa C, Fanetti G, Masello LC, Poli G (1988) Improvement of *in vitro* natural antibacterial activity in HIV infected patients (Submitted for publication)

Tagliabue A, Nencioni L, Villa L, Keren DF, Lowell GH, Boraschi D (1983) Antibody-dependent cell-mediated antibacterial activity of intestinal lymphocytes with secretory IgA. Nature 306:184-186

Tagliabue A, Villa L, Boraschi D, Peri G, De Gori V, Nencioni L (1985b) Natural antibacterial activity against *Salmonella typhi* by human $T4^+$ lymphocytes armed with IgA antibodies. J Immunol 135:4178-4182

Tagliabue A, Villa L, De Magistris MT, Romano M, Silvestri S, Boraschi D, Nencioni L (1986) IgA-driven T cell-mediated antibacterial immunity in man after live oral Ty21a vaccine. J Immunol 137:1504-1510

Wahdan MJ, Serie C, Cerisier Y, Sallam S, Germanier R (1982) A controlled field trial of live *Salmonella typhi* strain Ty21a oral vaccine against typhoid: Three year results. J Inf Dis 145:292-296

Weber J (1986) Gastrointestinal disease in AIDS. Clin Immunol Allergy 6:519-541

Yugoslav Typhoid Commission (1964) A controlled field trial of the effectiveness of acetone-dried and inactivated and heat-phenolinactivated typhoid vaccines in Yugoslavia. Bull WHO 30:623-630

Oral Immunization: A Summary

J.R. McGhee, and J. Mestecky

Department of Microbiology, University of Alabama at Birmingham, UAB Station, Birmingham, AL 35294, USA

Papers presented at this conference convincingly demonstrate that oral immunization represents an attractive and feasible alternative to systemic immunization routes particularly for the induction of specific antibodies of the IgA isotype in various external secretions. Recent technical improvement in the antigen delivery systems should further facilitate the acceptance of oral immunization as a route of choice for the prevention of mucosa-contracted diseases of viral and bacterial etiology as well as allergies.

Antigen Delivery Systems

In comparison with systemic immunization, induction of the effective mucosal immune response by the oral route generally requires doses of antigen that may be higher by several orders of magnitude due to the digestion of antigens by the gastrointestinal enzymes and their limited absorption from the gastrointestinal tract. From the standpoint of mucosal adjuvants and antigen delivery systems, several novel approaches are under development. antigen packaging into liposomes, as discussed by Michalek et al. or biodegradable microcapsules (Eldridge et al.) have yielded promising results. In animals, both types of vaccine formulation have been shown to enter Peyer's patches via M cells and initiate secretory and/or systemic immune responses. Even more important is the finding that microencapsulation of antigens into optimally sized particles that remain in the Peyer's patches for a desired time period (depending on the chemical composition of biodegradable material) would induce a prolonged immune response due to the continued stimulation of resident B and T cells. Thus, the frequently encountered problem of the inheritantly short IgA immune response memory may be overcome by this interesting approach.

Chemical conjugation of antigens to cholera toxin or its B subunit offers an alternative possibility (Elson et al.; Holmgren et al.). The high affinity of cholera toxin/B subunit to GM1 gangliosides on intestinal cells would assure a greater penetration via receptor-mediated endocytosis of conjugated antigens and consequently induce a vigorous intestinal as well as systemic immune responses. Although coupling of cholera toxin/B subunit to many different antigens is possible, the induction of an immune response to this general "carrier" may ultimately interfere with its binding to the epithelial cells and thus limit the long term usage.

Another interesting approach in the development of novel vaccines is based on the ability of auxotrophic strains of *Salmonella* to colonize the gastrointestinal tract and particularly Peyer's patches. Several *Salmonella* mutants have been generated in many laboratories that are avirulent or display diminished virulence and can be safely used in animals and hopefully in humans. The insertion of plasmids containing genes for virulence factors of other microorganisms renders such genetically modified

Current Topics in Microbiology and Immunology, Vol. 146

Salmonellae premier candidates for a new type of recombinant vaccines. Curtiss et al. illustrated this approach in this symposium. however, the use of carrier strains has a potential drawback in that secretory antibodies induced by a repeated exposure to a given vaccine strain may prevent its attachment and penetration into the Peyer's patch and thus greatly reduce the magnitude of the immune response due to limited delivery of desired plasmid-coded antigen into the gut-associated lymphoid tissue.

Combination or sequential use of the approaches outlined above, for example microencapsulation of antigens with cholera toxin/B subunit, may overcome certain limitations of each of the novel antigen delivery system designed for an optimal and long lasting humoral immune responses in external secretions as well as serum. Much experimental work will be necessary to address other related problems. The potential efficiency of such novel vaccines given by the oral route for the prevention of infections or allergic diseases of the respiratory or genitourinary tracts deserves further studies. Oral immunization followed by a topical application of antigens may be necessary to maximize the immune response in, for example, the upper respiratory tract. Furthermore, the frequency of immunization, combinations of the routes of delivery, and packaging may vary for each antigen used in the targeted induction of an immune response in various mucosal compartments or for a generalized systemic immunity.

Oral Immunization with Viral Vaccines

There is some evidence suggesting that oral vaccines that contain whole killed or genetically modified viruses or their component antigens, induce an immune response in external secretions and/or serum that is protective in some experimental systems. Chen, Bergmann, Waldman, Lamm, Koprowski, Moldoveanu, and their co-workers presented interesting results concerning this subject. Further studies of suitable antigen delivery systems of selected viral antigens will be necessary to develop new vaccines for the prevention of viral diseases afflicting the upper respiratory or gastrointestinal tracts such as infections with influenza, parainfluenza, respiratory syncithial, rabies, adeno, reo, and rota viruses.

In most cases the immunity to viruses is assessed by measuring the titers of specific antibodies in external secretions and/or in serum. However, the effective anti-viral immunity may not correlate with the level of antibodies in various compartments since immunized animals may be protected and yet display low antibody levels. In such cases, T cell-mediated immunity may be of considerable importance especially in the phase of recovery from a disease. Our current knowledge of the induction of T cell-mediated immunity induced by the oral immunization route is extremely limited. We know that immunization with viral nuclear proteins can induce T helper (Th) cells and these cloned cells can provide help for antibody responses to surface viral components if linked recognition is used. For example, clones recognizing nuclear proteins of influenza virus can provide Th function for anti-hemagglutinin responses. Of course, cytotoxic T cells can provide good class I-restricted responses in mice immunized with viral nucleoprotein. One could predict that this may be the main type of immunity induced by oral viral vaccines since internal viral components would be exposed following extensive enzymatic cleavage in the small intestine. Th cells may also selectively support IgA responses; however, the IgA may be directed to internal proteins and, thus, would be less effective in terms of viral neutralization.

Elegant work of the Waldman and Bergmann groups has shown that oral immunization with the influenza vaccine induces secretory IgA but not serum responses. These studies have been extended to important health groups, i.e., the elderly individuals.

Chen and associates have demonstrated that oral immunization of mice with influenza induces precursor IgA B cells for hemagglutinin and these B cells migrate from the gut-associated lymphoid tissues to mucosae of the upper respiratory tract where differentiation to IgA anti-hemagglutinin-producing cells occur. Lamm and co-workers demonstrated in mice orally immunized with Sendai virus vaccine how one can advantageously use the common mucosal immune system. Repeated oral immunization with Sendai induces S-IgA responses in the gut, but poor lung responses, even when boosted with cholera toxin. However, intranasal boosting after gut priming resulted in lung antibody responses, and a protective immunity from challenge with live Sendai Virus. Interestingly, the protective responses in the upper respiratory tract were mediated by secretory IgA while those in the lung were of the IgG class.

Bacterial Oral Vaccines

Studies in this area continue to advance rapidly and several vaccines have been used in humans (as described by Holmgren, Clancy, Taubman, Tagliabue, and their co-workers) and in animals (Hanson, Keren, Hale, and Michalek et al.). Holmgren and his associates have devoted intense efforts to the development of novel oral vaccines to *Vibrio cholerae, Shigella,* and enteropathogenic *Escherichia coli.* Continued success in the induction of protective immunity deserves further studies of the underlying mechanisms responsible for their immunity as well as longevity of the protection induced by the oral vaccines. Tagliabue and co-workers explored some of the protective mechanisms, particularly those involving the combined effector functions of IgA and T lymphocytes, and described thought-provoking and intriguing results. Further studies of the molecular and cellular parameters should allow detailed definition of the interaction of various forms of IgA with cell populations engaged in the antibody-dependent cell-mediated immunity.

Keren and Hale, with their colleagues, have made excellent progress with recombinant strains carrying *Shigella* plasmids. These investigators have shown convincingly that plasmids introduced into *E. coli* or *Salmonella* and given to experimental animals by the oral or local routes result in excellent S-IgA anti-shigella responses. The Thiry-Vella loop rabbit model continues to be useful and informative for evaluation of protective immunity and memory induction to these gram negative bacteria.

Clancy with his co-workers continue to provide interesting data concerning the successful induction of humoral immune responses in the respiratory tract of patients with bronchitis by oral immunization with nontypable *Haemophilus influenzae.* These studies may have clinically important implications for the prevention of episodes of seasonal bronchitis in certain patient populations. Future studies of the subclasses of IgA antibodies to individual antigens of *H. influenzae,* such as protein antigens, LPS or IgA1 protease frequently produced by upper respiratory tract pathogens could provide information essential for a rational design of orally administered vaccines against infections of the respiratory tract. Studies of Taubman et al. performed in humans and Michalek et al. carried out in an animal model, convincingly demonstrate that an oral vaccine for the prevention of dental caries represent an attainable goal. Purified antigens of mutans streptococci in liposomes or other adjuvants in oral vaccines given to experimental animals confer a significant protection. Extensive studies performed by Taubman's group in children and young adults have provided convincing evidence for an early maturation of the secretory immune system and its ability to respond promptly to an oral challenge with vaccines containing purified glucosyltransferases of mutans streptococci. Furthermore, our long lasting apprehension with respect to the induction of heart cross-reactive antibodies as a consequence of immunization with *Streptococcus mutans* has become less acute. Recent well-controlled immunochemical and immunohistochemical studies have suggested that the earlier observations may have been overinterpreted and rekindle our enthusiasm for the development of dental caries vaccines.

Humoral immune responses in external secretions may also be induced by systemic immunization, especially as a consequence of *prior exposure* to the same microorganism through the mucosal surfaces. Thus, Tarkowski and co-workers have demonstrated that systemic immunization with commercial vaccines containing polysaccharide antigens induce both serum and secretory antibodies, predominately of IgA2 subclass. The induction of protective antibodies to carbohydrate antigens remains a problem, especially in young children. Robbins, Schneerson and their colleagues have presented impressive data with protein-carbohydrate conjugate vaccines to enhance the responsiveness in this selected population. Further analyses of immune responses with respect to serum and secretory antibodies and IgG and IgA subclass association of specific anti-protein and anti-carbohydrate antibodies should yield additional information essential for design of effective vaccines.

Hanson's work raised several intriguing questions. The idea of maturation of avidity and differences in the kinetics of IgA antibodies upon repeated immunizations with lipopolysaccharide versus protein antigens will undoubtedly stimulate considerable research particularly in the area of appropriate antigen presentation and T cell regulation of such responses.

Future Studies

It is safe to conclude that the common mucosal immune system can be exploited to the host's advantage in the prevention of numerous infections and allergic diseases. Although many important aspects such as optimal doses and frequency of immunizations and most effective antigen delivery system, deserve intensive explorations, progress that has been attained warrants optimism. However, it is important to think beyond the common mucosal immune system and to consider other areas of related research. For example, the effector function of intraepithelial lymphocytes and other T cell populations which may function in concert with mucosal antibodies of various isotypes deserve due attention. We must start to think about the origin, migration, and homing of these large T cell populations and their effector functions in anti-viral, anti-bacterial, and even anti-parasitic immunity.

Another area of future productive research concerns the problem of antigen presentation and antigen processing by the mucosa-associated tissues. Epithelial and other accessory cells present in these locales express, particularly under the influence of various lymphokines, class II antigens. This may constitute *in vitro* as well as *in vivo* an essential step in an effective antigen presentation and/or processing for induction of humoral and cell-mediated immune responses or tolerance to a broad spectrum and quantity of environmental antigen. Furthermore, accessory cells as well as T cell and their corresponding lymphokines such as interleukins 4, 5, and 6 may have a profound impact on the regulation of the immunoglobulin isotype expression by resident B cells found in mucosal tissues at various stages of their differentiation. This, too, may have important implications with respect to the generation of local IgA and IgG versus IgE immune responses and their clinical consequences. All of these and other ancillary problems represent a challenge for continued research in this important area. We should realize with increasing acuity, the importance and relevance of our efforts for the prevention of severe diarrhoeal, respiratory, and venereal diseases, including AIDS, that afflict yearly millions of individuals and are frequent causes of mortality. We think that oral immunization may, due to the simplicity and acceptability of administration, significantly improve our chances in eliminating or alleviating these diseases.

MIX
Papier aus verantwortungsvollen Quellen
Paper from responsible sources
FSC® C105338

If you have any concerns about our products,
you can contact us on
ProductSafety@springernature.com

In case Publisher is established outside the EU,
the EU authorized representative is:
Springer Nature Customer Service Center GmbH
Europaplatz 3, 69115 Heidelberg, Germany

Printed by Libri Plureos GmbH
in Hamburg, Germany